Adaptation to Climate

The impacts of climate change are already being felt. Learning how to live with these impacts is a priority for human development. In this context, it is too easy to see adaptation as a narrowly defensive task – protecting core assets or functions from the risks of climate change. A more profound engagement, which sees climate change risks as a product and driver of social as well as natural systems, and their interaction, is called for.

Adaptation to Climate Change argues that without care, adaptive actions can deny the deeper political and cultural roots that call for significant change in social and political relations if human vulnerability to climate change associated risk is to be reduced. This book presents a framework for making sense of the range of choices facing humanity, structured around resilience (stability), transition (incremental social change and the exercising of existing rights) and transformation (new rights claims and changes in political regimes). The resilience–transition–transformation framework is supported by three detailed case study chapters. These also illustrate the diversity of contexts in which adaption is unfolding, from organisations to urban governance and the national polity.

This text is the first comprehensive analysis of the social dimensions to climate change adaptation. Clearly written in an engaging style, it provides detailed theoretical and empirical chapters and serves as an invaluable reference for undergraduate and postgraduate students interested in climate change, geography and development studies.

Mark Pelling is a Reader in Geography at King's College London and before this at the University of Liverpool and University of Guyana. His research and teaching focus on human vulnerability and adaptation to natural hazards and climate change. He has served as a lead author with the IPCC and as a consultant for UNDP, DFID and UN-HABITAT.

Adaptation to Climate Change

From resilience to transformation

Mark Pelling

Routledge
Taylor & Francis Group

LONDON AND NEW YORK

First published 2011
by Routledge
2 Park Square, Milton Park, Abingdon, Oxon, OX14 4RN

Simultaneously published in the USA and Canada
by Routledge
270 Madison Avenue, New York, NY 10016

Routledge is an imprint of the Taylor & Francis Group, an informa business

Typeset in Times New Roman
by Keystroke, Tettenhall, Wolverhampton
Printed and bound in Great Britain
by TJ International Ltd, Padstow, Cornwall

British Library Cataloguing in Publication Data
A catalogue record for this book is available from the British Library

Library of Congress Cataloguing in Publication Data
Pelling, Mark, 1967-
Adaptation to climate change / Mark Pelling.
p. cm.
Includes bibliographical references and index.
ISBN 978-0-415-47750-5 (hardback) — ISBN 978-0-415-47751-2 (pbk.)
1. Climatic changes. 2. Climate change mitigation.
3. Human beings—Effect of climate on. 4. Acclimatization. I. Title.
QC903.P44 2010
304.2′5—dc22
2010013609

ISBN: 978–0–415–47750–5 (hbk)
ISBN: 978–0–415–47751–2 (pbk)
ISBN: 978–0–203–88904–6 (ebk)

For Ulli and Lilly

Contents

List of illustrations

List of acronyms and abbreviations

ASEAN	Association of Southeast Asian Nations
CARE	Cooperative for Assistance and Relief Everywhere (*formerly the* Cooperative for American Remittances to Europe)
CBA	cost benefit analysis
CHS	Commission on Human Security
COP	Conference of the Parties
DEFRA	Department for Environment, Food and Rural Affairs (UK)
DETR	Department of the Environment, Transport and the Regions (UK)
DFID	Department for International Development (UK)
DOE	Department of the Environment (UK)
EIU	Economist Intelligence Unit
FAO	UN Food and Agriculture Organisation
FEMA	Federal Emergency Management Agency (USA)
GDP	gross domestic product
GECHS	Global Environmental Change and Human Security Programme
GNAW	Government of the National Assembly of Wales
GROOTS	Grassroots Organisations Operating Together in Sisterhood
IADB	InterAmerican Development Bank
IDESO	Universidad Centroamericana, Instituto de Encuestas y Sondeos deOpinión
IFRC	International Federation of the Red Cross and Red Crescent Societies
INETER	Institute for Territorial Studies (Nicaragua)
IPCC	Intergovernmental Panel on Climate Change
ISDR	UN International Strategy for Disaster Reduction
MAFF	Ministry of Agriculture Fishers and Food (UK)
NGO	non-governmental organisation
ODA	overseas development assistance
OECD	Organisation of Economic Cooperation and Development
SDN	Sustainable Development Networking

SES	socio-ecological system
SEDESOL	Secretaría de Desarrollo Social (Mexico)
SEMARNAT	Secretaría de Medio Ambiente y Recursos Naturales (Mexico)
SINAPRED	Sistema Nacional Para la Protectión, Mitigación y Atención de Desastres (Nicaragua)
UK	United Kingdom
UN	United Nations
UNCED	United Nations Conference on Environment and Development
UNHABITAT	United Nations Human Settlements Programme
UNDP	United Nations Development Programme
UNFCCC	United Nations Framework Convention on Climate Change
USA	United States of America
USAID	United States Agency for International Development
USGS	United States Geological Survey
WRI	World Resources Institute (USA)
WWF	Worldwide Fund for Nature

Acknowledgements

This book would not have been possible without the inspiration and generous exchange of ideas with colleagues, in particular: Kathleen Dill (Cornell University), Cris High (Open University), David Manuel-Navarrete (King's College London) and Michael Redclift (King's College London). Research underpinning this book was undertaken as part of three grants awarded by the UK Economic and Social Research Council (RES-221-25-0044-A, RES-228-25-0014 and RES-062-23-0367). Without this financial support the work would not have been possible. I would also like to give thanks for the many rich discussions I've enjoyed with PhD and masters students in the Department of Geography at King's College London – two PhD graduates, Marco Grasso and Llewellyn Leonard, have their theses referenced. Wider discussions, in particular through the Global Environmental Change and Human Security Programme led by Karen O'Brien at the University of Oslo, have also been instrumental in shaping this work. For patience beyond the call of duty I thank the Routledge editorial and print teams, in particular Andrew Mould. Special recognition is also due to Ulli Huber for creating the time and atmosphere necessary to complete this work and to Lilly Pelling for her word processing and computer management skills. Of course, the real thanks go to all those respondents who have freely given of their time and energies to provide the empirical backbone for this work, many of whom remain at the sharp end of adapting to the consequences of climate change and development failure.

Part I
Framework and theory

1   The adaptation age

Everyone has the right to life, liberty, and security of person.
(Universal Declaration of Human Rights, Article 3)

Climate change adaptation is an opportunity for social reform, for the questioning of values that drive inequalities in development and our unsustainable relationship with the environment. But this outcome is by no means certain and growing evidence suggests that too often adaptation is imagined as a non-political, technological domain and enacted in a defensive rather than a progressive spirit. Adaptation has been framed in terms of identifying what is to be preserved and what is expendable, rather than what can be reformed or gained. Dominant development discourses put the economy as first to be preserved, above cultural flourishing or ecological health. There is a danger that adaptation policy and practice will be reduced to seeking the preservation of an economic core, rather than allowing it to foster the flourishing of cultural and social as well as economic development, or of improved governance that seeks to incorporate the interests of future generations, non-human entities and the marginalised.

The argument put forward in this book suggests that adaptation is a social and political act; one intimately linked to contemporary, and with the possibility for re-shaping future, power relations in society. But it also recognises that different actors perceive contrasting roles for adaptation. That there may be multiple ways of adapting is already recognised in the literature through the range of different scopes and timings for adaptive interventions (for example, Smit *et al.*, 2000; Smit and Wandel, 2006). These are important technical considerations but more emphasis is needed on the underlying socio-political choices that are made through the selection of adaptation pathways. Here we propose three such pathways leading to resilience (maintaining the status quo), transition (incremental change) and transformation (radical change). No one pathway necessarily leads to 'progressive' or more equitable and efficient outcomes than the others. The evaluation of pathways and subsequent outcomes will be a function of context and the viewpoint of individual actors. Opening analysis of how it is that individual adaptive pathways come to dominate or be marginalised is one of the aims of this book, which offers theoretical and empirical exploration.

Recent experience suggests that consensus on a progressive adaptation will not be easy. Our current age of adaptation is the second time in recent history that a global environmental challenge has provided an opportunity to question dominant forms of development. The first, coalescing around the notion of sustainable development, has (to date) manifestly failed. The international roots of the sustainable development agenda lay in a concern that the environmental limits to economic growth were fast approaching. Indeed the combination of mitigation and adaptation agendas represents a reprise of the sustainable development agenda, and climate change a strong signal that existing developments are far from sustainable (Le Blanc, 2009). Underlining the significance of adaptation for sustainable development, Adger *et al.* (2009a) remind us that climate change adaptation decisions have justice consequences across as well as within generations.

The first mainstream expression of a sustainable development approach was the Brundtland Commission, 1983, which stimulated a search for radical ecological and social alternatives to development (Redclift, 1987). These peaked in public awareness at the UN Conference on Environment and Development (UNCED) in Rio de Janeiro, Brazil, 1992. Here, differences in the prioritising of development and environment between rich and poorer nations and the influence of a strong industry lobby limited reform at the international level. The parallels with current challenges facing international negotiations at the UNFCCC are striking. The policy legacy of UNCED has been a constrained version of sustainable development largely restricted to ecological modernisation and an acceptance of the substitutability of environmental for economic value (Pelling, 2007a). Where some success has been achieved through this process it is outside of the compromised domain of international politics through the innovations of civil society groups, fair trade companies and concerned individuals, where environmental and social justice goals have been brought into projects for economic development (Adams, 2008). But these initiatives remain fragmented and overwhelmed by the global policy consensus.

Can climate change adaptation reinvigorate these debates and provide an impetus for stronger sustainable development action? Might climate change adaptation be both a reprise of sustainable development and a new opportunity in its own right? The origins of the UNFCCC process lie partly in UNCED where the first Framework Convention on Climate Change agreement was opened for signature. This connection to debates on sustainable development also reminds us that climate change and resultant adaptation are but one expression of an underlying crisis in environment–society relationships. The deepest root causes of climate change and the inability of those with power in society (locally and globally) to act lie in the dominant processes and values of the political economy that increasingly concentrate wealth in the hands of a few, with unjust social and environmental externalities as accepted. At this level climate change risk is but one expression of a deeper social malaise in modern society. For the poor, comfortable and rich alike aspiring and acquiring in order to consume have become the rationale for development; a rationale propelled as much by fear of failure as the pleasures of consumption. Can the burgeoning academic and policy interest in adaptation be levers to address these deeper questions of sustainability and justice, as well as adjusting to meet the more proximate risks presented to us by a changing climate?

Here we propose and illustrate a framework to help reveal and understand the social, cultural and political pathways through which adaptation to climate change unfolds. Adaptation is conceptualised through three layers of analysis (Chapters 3–5) which build from a starting point in the notion of resilience to encompass adaptation as a process of socio-political transition and transformation. Each stage of theoretical analysis brings together work from systems theory with a wider literature including regime analysis, discourse, risk society, human security and the social contract. This reflects the strong influence of systems thinking on adaptation work but also enables the theoretical precision derived from systems thinking – for example, on social learning and self-organisation – to run throughout the book while bringing to the fore power, which is more ably addressed through other theoretical discourses. These theoretical discussions are then illustrated through three case study chapters showing how adaptation can unfold through contested politics in organisations, urban systems and nation states.

Power lies at the heart of this conceptualisation of adaptation. Power asymmetries determine for whom, where and when the impacts of climate change are felt, and the scope for recovery. The power held by an actor in a social system, translated into a stake for upholding the status quo, also plays a great role in shaping an actor's support or resistance towards adaptation or the building of adaptive capacity when this has implications for change in social, economic, cultural or political relations, or in the ways natural assets are viewed and used. Accepting that adaptation is contested makes interpreting adaptation as progressive hostage to the observer's viewpoint. This requires the imposition of a normative framework to provide a consistent and transparent positionality for analysis. Here we are guided by Rawls' theory of justice that identifies procedural (inclusion in decision-making) and distributional (social and spatial) elements. Rawls (1971, see also Paavola *et al.*, 2006) prioritises human rights over public goods; holds the social contract between citizens and the state in dynamic tension so that it is liable to capture by vested interests at moments of pressure; and argues that society should be governed by principles that protect inclusive governance and seek to enhance the quality of life of the poorest. This final statement is perhaps the most important for making judgements on comparative adaptation pathways.

In seeking to make the social and political elements of adaptation visible three questions run throughout this book and structure its narrative:

1. How is adaptive capacity shaped?

    Or, to what extent is adaptive capacity dependent upon existing institutional and actor capacity; can it be constructed anew through external influence or through autonomous actions?

2.   How is adaptive capacity turned into adaptive action?

Or, what institutions and actors are important in mediating this threshold and the wider feedback between action and future capacity?

3.   What are the human security outcomes of adaptive actions?

Or, how far do framing institutions and individual actors control processes of adaptation and how does this affect the exercise of rights and responsibilities in society, and the social distribution of well being, basic needs, human rights and subsequent adaptive capacity?

The following sections in this introductory chapter establish the scope of the book. First adaptation is defined and the approach taken to make climate change and associated adaptations visible explained. Second, to help contextualise this work, some of the main strands in contemporary adaptation theory are presented. Finally the structure of the book is outlined.

Adapting to climate change

Adaptation in the face of environmental change is nothing new. Individuals and socio-ecological systems have always responded to external pressures. But climate change brings a particular challenge. Uncertainty in the ways through which climate change will be felt set against its speed and scale of impact, combined with the invisibility of causal linkages in everyday life, bring new challenges for the sustainability of socio-ecological systems. It is for this reason that understanding adaptation to climate change is a critical challenge of our time. As the title of this book suggests, adaptation is conceived of here as a dynamic phenomenon – as a process rather than a status. An individual or business may be well adapted to a particular moment in history, but the dynamism of climate change requires an adaptation that can coevolve with it. Climate change is no longer an external threat to be managed 'out there', but is an intimate element of human history – both an outcome and driver of development decisions for individuals, organisations and governments. This requires a closer look at social relations and practices, even values, as sites for adaptation, and suggests that it is necessary, but not sufficient, to control the impacts of climate change through technological innovations like environmental engineering and crop selection.

There are many ways of characterising adaptation, which as an intellectual construct cannot be directly observed. Here a key distinction is made between adaptation that is forward or backward looking. As a backward looking attribute, adaptation is revealed by capacity to cope during moments of stress or shock. For example, well-adapted urban communities have fewer losses to hurricane events. Greater capacity in Cuba's early warning and evacuation systems when compared to the southern states of the USA in large part explain the far lower human losses in Cuba from hurricane events (UNDP, 2004). As a forward looking attribute, adaptation cannot be revealed through impacts (which have not yet happened) and instead is made visible through theoretically identified components associated with adaptive capacity. An important gap in our understanding of adaptation comes from the difficulty of being able to follow adaptive processes over time and so verify through observation the contribution of theoretically defined components on adaptive practices.

Despite this caveat, our focus is on forward looking adaptation. It is here that adaptation has the potential to intervene in development policy and practice through progressive risk reduction. To this extent the work is driven by theoretical understandings of what constitutes adaptive capacity. On the ground, however, past experiences that reveal backward looking adaptation can feed in to local understandings of the pressures shaping capacity looking forward. A full discussion of adaptation theory is presented in Chapter 2.

For researchers and policy makers alike the invisibility of forward looking adaptive capacity is compounded by the dynamism of climate change. For specific physical or ecological systems change can be gradual and persistent – for example, in sea level rise. For others temporary equilibrium may be violently disrupted when thresholds are breached and systems enter new states – for example, the potential reversal of the thermohaline circulation system in the North Atlantic. The impact of such global scale processes is mediated by local socio-ecological and environmental conditions. This has led many to argue that adaptation is a local agenda in contrast to mitigation, which is global. While our concern is with adaptation, we make a case for both agendas to have local and global components and indeed national level action too. High level legal frameworks and voluntary agreements can support local action, but local level action is also a potential driver for higher level policy. Where political will is absent at higher levels, local action has the potential to be decisive in determining capacity and action and influencing higher level policy. This is the case for mitigation and adaptation – for investing in zero carbon lifestyles and technology as much as livelihood diversification. Those fundamental social attributes that enable and shape adaptive capacity also influence the potential for local contributions to mitigation (Bulkeley and Betsill, 2003).

Climate change is also a slippery concept to demonstrate empirically. Outside of the imaginary worlds of computer models it is as yet impossible to determine the proportion of any hydrological or meteorological event that is attributable to climate change. O'Brien and Leichenko (2003) were among the first to argue that searching for the incremental risk associated with climate change is a lost cause and many years away from resolution. Meanwhile the numbers of people and socio-ecological systems at risk and bearing loss from climate change associated events is increasing. Climate change is manifest locally through extreme events and in the heightened variability of precipitation, temperature and wind. We may never understand the precise contribution of anthropocentric climate change to these events and trends but we can be certain that climate change is a decisive contributing factor and that vulnerability exists, demanding action.

The idea of adaptation

While mitigation was clearly defined in the original United Nations Framework Convention on Climate Change (UNFCCC) negotiated in the Rio Summit, 1992, adaptation was not. Despite this the term was used in the agreement text. Its meaning continues to be debated (Burton, 2004). Arguably it is the slipperiness of the term that has been part of its attraction for discussion in academic and policy circles alike. Here we present an overview of some of the main contributions to the adaptation debate as scholars and policy makers have sought to make sense of the term handed to them from the UNFCCC. The section begins with an assessment of the influence of the IPCC–UNFCCC on scholarly work on climate change adaptation, of the ways in which climate change impacts are evaluated and the geographical distribution of climate change impact risk. From this point an overview of work on social aspects of adaptation is presented around four questions that cross-cut research. This discussion is a prelude to that in Chapter 2, which offers an extended response to the intellectual inheritance and current shape of adaptation to build a conceptual framework.

The IPCC–UNFCCC frame

The IPCC and UNFCCC procedures and agendas have greatly influenced the direction of thinking as well as policy on climate change adaptation. There is a high level of interaction between these institutions, with the IPCC feeding into the UNFCCC process, which in turn helps to drive funding and political will for adaptive actions and research. The stated aim of the IPCC is to support national policy on climate change through offering scientific consensus. Founded in 1988, the IPCC has produced Assessment Reports in 1990, 1995, 2001 and 2007. Each in turn has included a greater emphasis on adaptation as evidence has accumulated. In this way the IPCC has acted as both a stimulus and a resource for research on adaptation to climate change.

The First Assessment Report helped to shape the UNFCCC and drive its ratification at the UN Conference on Environment and Development, Rio, 1992 (Agrawala, 2005), but said relatively little about adaptation. It was in the Second Assessment Report that the socio-economic aspects of climate change were seriously addressed for the first time. The report concluded by sketching out the scope of support needed for adaptation. It argued that efficient adaptation depended upon the availability of financial resources, technology transfer and cultural, educational, managerial, institutional, legal and regulatory practices, both domestic and international. The vision was firmly on the potential roles and responsibilities of international actors with limited evidence of local adaptive behaviour. The Third Assessment Report included a greater focus on adaptation strategies and concluded that adaptation was necessary to complement mitigation efforts raising the significance of adaptation in the UNFCCC process and helping to achieve the Nairobi work programme. The Fourth Assessment Report stated that adaptation was necessary to address the impacts of climate change, was clear that this was already occurring and that more extensive adaptation than was being undertaken would be necessary to address future vulnerability to climate change. Hinting at the possibility of a progressive adaptation agenda, the report also connected sustainable development with vulnerability to climate change, and argued that climate change could impede national abilities to follow sustainable development pathways. This report provided the scientific basis for the Bali Action Plan reached by parties at the 13th session of the Conference of the Parties (COP).

While the IPCC process is scientific it reports back to governments and is influenced by their interests and priorities (Grundmann, 2007). It has been described as a boundary object, and a hybrid science-policy project at the interface between science and politics. As a consensus organisation, and one open to intense public scrutiny, it is conservative, careful to follow core science rather than policy or advocacy trends. This has been its strength in terms of scientific credibility for policy makers, but has also made it difficult for some evidence on climate change and human reaction to be included. For example, much local evidence for climate change impacts and experience in adaptation, particularly in Africa, Asia and Latin America and the Caribbean, is gained by local actors and held by civil society actors or published nationally or regionally – but not in international peer review journals – and so has been difficult to include. The IPCC in this respect offers a conservative, rigorous view of climate change, but should not be seen as a full acccount of existing information or knowledge. The Fourth Assessment Report from Working Group II began to address this by drawing also from the grey literature produced by governments and NGOs. A concern for inclusiveness in scientific representation from all world regions has also led to a quota system and travel funds to support participation from scientists based in low- and middle-income countries. Even so, this does not mean that governments are equally happy with IPCC findings, with various US governments largely ignoring the IPCC while others (especially in Europe) have endorsed and acted upon it through the UNFCCC and unilaterally (Grundmann, 2007).

The IPCC process has also been constrained by its slow recognition of the full contribution to climate change debates to be made from parallel disciplines or policy areas that may cover very similar ground but not use the language of climate change or publish in climate change associated journals. Thus, for example, the considerable academic and policy literatures on disaster risk reduction, social security and food security in developing countries, community-based water management and risk insurance are making only slow impact on the IPCC. Such a sharp focus was perhaps appropriate to managing information on the natural and physical science components of climate change. For Working Group II's remit of vulnerability and adaptation, where impacts and responses often build on past experience but ultimately transcend policy or disciplinary boundaries, this is less helpful – at times threatening that the IPCC will reinvent theoretical or methodological lessons that could better be brought in from other specialisms.

For example, much of early conceptual work on adaptation mirrored existing work on coping within the food security and disaster risk disciplines. The Fourth

Assessment Report from Working Group II went some way to addressing this concern with the inclusion of cross-sector and indeed cross-report case studies including the consequences and responses to Hurricane Katrina and the vulnerability of mega deltas (IPCC, 2007). Working Group II also took the lead role between 2009–11 in organising a Special Report on Managing the Risks of Extreme Events and Disasters to Advance Climate Change Adaptation, to help bring knowledge from disaster risk management into climate change adaptation. At this stage of mapping and understanding vulnerability and adaptation the inclusion of knowledge from cognate areas is important. However, it is also key that the IPCC be seen in context. It is a mechanism for consolidating knowledge for the policy community on climate change. As the IPCC increasingly recognises parallel communities the challenge will be in retaining its core purpose and intellectual focus while embracing ever wider sources of knowledge. For the academic community the challenge is to communicate effectively with the IPCC process without restricting analysis and thought to the priorities of the IPCC.

The costs of adapting

It is difficult to estimate the future costs of adapting to climate change. From the more restricted world of disaster management we know that the difference between investing in prevention and the costs of a disaster impact can easily exceed a ratio of 7:1 (DFID, 2004a). The costs of adapting to climate change are more far-reaching. Future demands for adaptation are also to be shaped by the actions we take now to mitigate it. One estimate, by Stern (2006), suggests adaptation costs in the order of 5 to 20 times the estimated costs of containing climate change through mitigation. The economic costs of adaptation are also not evenly distributed worldwide. Using data from past natural disaster events shows that richer nations with an accumulated legacy of physical infrastructure and housing have the most absolute economic exposure. However, these countries also have the assets to adapt (and arguably have substantial power and duty to mitigate, and in this sense have some control over, their own destiny). Poorer countries have less physical assets exposed but economies tend to be more dependent upon primary production and ecosystem services. As a proportion of GDP potential economic losses are highest in low and middle income countries. In addition, it is the same countries – from Africa, Central, South and Southeast Asia and Central America and the Caribbean – that record the highest mortality rates from natural disasters, adding human to economic vulnerability (UNDP, 2004). Past experience and projected risk of human loss through mortality and morbidity are also strongly skewed to poorer countries where income is dependent on primary extraction and where populations are not protected from environmental hazards by safe buildings, infrastructure, health services, and transparent and responsive governance (IFRC, 2010).

Observed data based on losses to past patterns of disaster events is the best guide to current vulnerability and backward looking adaptive capacity, but climate change means past patterns of hazard may not be as useful a guide to the future as had once been assumed; the so-called problem of non-stationarity

(Milly *et al.*, 2008). Figure 1.1 shows an example of forward looking assessment of relative vulnerability to climate change and extremes under a warming of 5.5 degrees C. It incorporates adaptive capacity as a component of vulnerability. Vulnerability is calculated based on input variables for human resources (dependency ratios and literacy rates), economic capacity (market GDP per capita and income distribution) and environmental capacity (population density, sulphur dioxide emissions, percentage of unmanaged land). The advantage of this approach is that it is not tied to past experiences of extreme events. Despite this, results largely confirm the burden identified above for poorer countries. High levels of vulnerability are associated with low and middle income countries in South America, southern Asia and Africa. High vulnerability is also found in China and some of Eastern Europe. But this method also suggests North America and Europe are extremely vulnerable, painting a portrait of widespread vulnerability across the globe where adaptive capacity is overwhelmed by climate change even over the next 40 years (Yohe *et al.*, 2006). This is a compelling case for the need for urgent and deep levels of mitigation alongside the need to support adaptation to reduce vulnerability from current and inescapable future climate variability and extremes.

The IPCC (2007) calculates that for the most exposed countries, such as coastal states in Africa, adaptation costs may be as high as 10 per cent of national GDP. For low-lying small island developing states the relative costs are even higher. Oxfam (2008) estimates that at least US$50bn is needed annually to support adaptation in developing countries. UNDP (2007) identifies an additional need of around US$86bn by 2015 (0.2 per cent of developed country GDP) on top of existing overseas development assistance budgets from bilateral and multilateral donors. These are large sums, but not unprecedented. The UNDP (2007) equates its total cost estimate to around 10 per cent of the current military expenditure by OECD countries.

The international architecture for support of adaptation is developing as adaptation rises on the political agenda. The UNFCCC provides one management structure for support of low and middle income countries. Bilateral and multilateral agencies, such as the development banks and other UN agencies, also provide financial and technical support. Investment decisions in the corporate private sector also impact on adaptation, including policy decisions from the insurance and reinsurance sectors, and are likely to increase in importance as businesses in middle and high income countries are forced to adapt. The emerging infrastructure is, however, built around existing poles of power – nation states and the UN system which is beholden to them, or banking interests, with nation states or private investors at the helm. Can these actors be expected to embrace adaptation as anything other than resilience – acts to reinforce the status quo? Indeed should they be encouraged to do so? Asserting more radical change in social and political systems needs to come from below through the actions of people at risk building on existing social and political reform movements.

With the costs of climate change increasing and adaptation being increasingly demanded, meeting the funding gap for adaptation in the short term is a key

Global distribution of vulnerability to climate change
Combined national indices of exposure and sensitivity

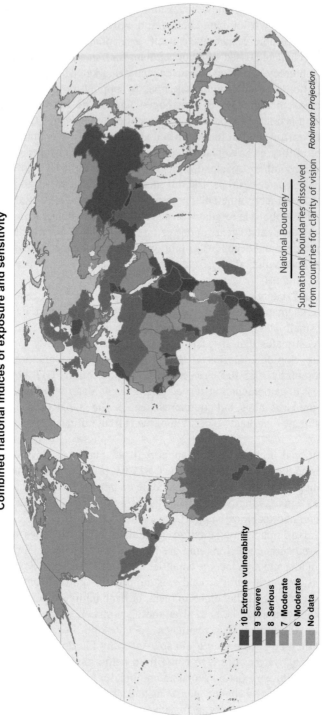

Scenario A2 year 2050 with climate sensitivity equal to 5.5 degrees C
Annual mean temperature with extreme events calibration

National Boundary—

Subnational boundaries dissolved
from countries for clarity of vision   *Robinson Projection*

10 Extreme vulnerability
9 Severe
8 Serious
7 Moderate
6 Moderate
No data

©2006 Wesleyan University and Columbia University

http://ciesin.columbia.edu/data/climate/

Figure 1.1   Global distribution of vulnerability to climate change. Combined national indices of exposure and sensitivity
(Source: Yohe *et al.*, 2006)

challenge. Without additional and earmarked funds for adaptation there is a risk of money being forced from existing overseas development assistance (ODA) budgets. ODA finance is already being squeezed by increased recent demand for humanitarian and disaster reconstruction funding (White *et al.*, 2004). Agrawala (2005) has estimated that between 15–60 per cent of official development assistance (ODA) flows will be affected by climate change. This trend is a particular tragedy as ODA is a key mechanism for reducing generic vulnerability to disaster risk and climate change impacts as well as achieving broader human security goals. A range of proposals exists for identifying additional funds. Oxfam (2008) proposes that funding be generated from auctioning a fraction of emissions allocations to developed countries under the post-2012 agreement, including proposed new emissions-trading for international aviation and shipping. Other proposals include increasing the share of the Clean Development Mechanism contributing to adaptation and increasing the role played by private capital through venture capital or commercial loans.

Conceptual development

Since its reintroduction into social scientific and policy debates following the Rio Summit, the interests of different analysis have made adaptation a slippery concept. For some, adaptation's contribution would best be as a tightly defined, technical term (like mitigation in the existing UNFCCC documentation) that can add universal clarity to policy formation including at the international level (for example, Schipper and Burton, 2009). Others, who see adaptation not as a technical category but as a research field, tend to have a wider view. Fankhauser (1998) suggests that adaptation can be synonymous with sustainable development. This challenge was noted as early as 1994 by Burton, just two years after the Rio Summit, and the plethora of interpretations has continued to grow as individual disciplines and intellectual communities have invested adaptation with their own worldviews (Kane and Yohe, 2007).

The adaptation to climate change debate is driven by four questions:

- What to adapt to?
- Who or what adapts?
- How does adaptation occur?
- What are the limits to adaptation?

None of these questions have easy answers.

What to adapt to?

Climate change itself is agreed to be manifest in at least three interacting and overlapping ways: climate change has come to encompass long-term trends in mean temperatures and other climatic norms, importantly precipitation, and secondary effects like sea-level rise together with variability about these norms from inter-seasonal to periods of a decade with particular implications for infrastructure planning, agriculture and human health, and extremes in variability that can trigger natural disasters such as floods, hurricanes, fires and so on (IPCC 2007). Furthermore, local studies of adaptation make it increasingly clear that while international and national policy makers may seek a clear measurement of impacts and adaptation associated with climate change – the incremental costs of mitigating or adapting to climate change, as the Global Environmental Facility puts it (Labbate, 2008) – on the ground, any meaningful measurement of adaptation needs to accept climate change is contextualised with the other risks (social, economic and political as well as environmental) that shape and limit human well being and the functioning of socio-ecological systems (Pelling and Wisner, 2009). This is the difference between an economic analysis of the farming sector of a country, and understanding the competing choices that shape adaptive capacity and actions for an individual farmer put in the context of the markets and regulatory regimes within which the farmer operates. Both are useful but partial lenses. The overlapping of seasonal and other climatic cycles with variation in baseline climate change and extremes makes it very difficult for specific events to separate out climate change signals from background weather patterns. Both short-term uncertainty in variability and extremes and long-term trends need to be considered (Adger and Brooks, 2003).

Who or what adapts?

Initial work on assessing who or what adapts came from the assessment of regional or national scale agro-economic systems. For example, Krankina *et al.* (1997) refer to boreal forestry management strategies as a means of assisting forests adapt. Here the system of interest was ecological and the management system an intervening variable between it and climate change. This kind of work complements well the scale of resolution available from climate modelling and the existing understanding of ecological adaptation within agricultural sciences, but is less suited to explore well the social processes driving and limiting adaptive decision-making. Economic assessment has also operated well at this scale, seeking to identify the costs (and benefits) of climate change scenarios for agricultural systems and to varying extents factoring in human adaptation. In a review of the economics of climate change literature, Stanton *et al.* (2008) observe the narrow framing used to approach decision making for climate change policy. Harvey (2010) goes further, arguing that a new macro-economic vision is needed to help move past the internal contradiction of contemporary economics that promotes energy intensive growth and so accelerates climate change with consequent growth inhibiting outcomes. Contemporary incentives push for greater and greater economic growth in an attempt to grow our way out of climate change and its attendant risks. The extraction and concentration of wealth that results increases collective vulnerability while simultaneously accelerating climate change associated (and other environments) hazards.

More human-centred analyses have also flourished which seek to identify the human and social characteristics that determine the capacity of communities to face a shock or stress (Adger *et al.*, 2005a). Local viewpoints help to contextualise adaptation within development and explain why people are unable or unwilling to take adaptive action (helping to identify the limits to climate change adaptation). From an analysis of two communities in Puerto Rico, Lopez-Marrero and Yarnal (2010) found that concerns for health conditions, family well being, economic factors and land tenure were given more priority by local actors than adaptation to climate change, despite their exposure to flooding and hurricanes. The results show the importance of addressing adaptation within the context of multiple risks, and of people's general well being.

How does adaptation occur?

The diversity of work examining processes of adaptation has benefited from a number of typologies of adaptive action and their coherent synthesis, see Smit *et al.* (2000), Smit and Wandel (2006), Burton *et al.* (2007). Carter *et al.* (1994) distinguished between autonomous (automatic, spontaneous or passive adaptations) that occur as part of the routine of a social system, and planned (strategic or active) adaptations. Smit *et al.* (2000) also add that adaptations may occur unintentionally as an incidental outcome of other actions – further emphasising the importance of contextualising assessments of adaptive capacity and action. The timing of the adaptation relative to its stimulus has led to additional types. Some draw from the disasters community, which uses a staged model of actions for tracking behaviour before and after disasters. Burton *et al.* (1993) distinguish adaptations that prevent loss, spread loss, change use or activity, change location or engage in restoration. More generally, adaptations can be reactive, concurrent (especially important for analysis of adaptation to gradual and ongoing changes in climatic norms) and anticipatory. Adaptive actions can be long- or short-term, and this has come to be associated with a distinction between actions aiming for short-term stability (coping) or longer term change (adaptation) (see Chapter 2). Adaptation has also been characterised according to the form of action (technological, behavioural, financial, institutional or informational), the actor of interest (individual, collection), the scale of the actor (local, national, international) and social sector (government, civil society, private sector); and the costs and ease of implementation (Smit *et al.*, 2000). Maladaptation is used to describe those acts that, through bad planning or inadvertent consequences, cause either local or distant consequences that outweigh gains (Smit, 1993).

What are the limits to adaptation?

Literature on the limits to adaptation has largely been framed by the concerns of international actors. The challenge so defined is to provide guidance for policy makers on what might be achieved through adaptation to limit or avoid the dangerous impact of climate change as a parallel agenda to mitigation – to help achieve a balance in investment between mitigation and adaptation (Hulme, 2009). At first this seems a technical problem of assessing the economic costs of a range of technical solutions and applying cost–benefit analysis to a range of scenarios. This can certainly help. But if adaptation is to move us towards a more sustainable development path then technological investments are only part of the solution. Changes in values and associated governance regimes will also need to be on the agenda, or may force themselves on as established institutions fail in the face of climatic extremes. Examining these limits is more difficult.

At root this approach to defining the limits to adaptation is contingent upon the levels of risk associated with climate change that are socially acceptable (Adger *et al.*, 2009a). Given the unequal social and geographical distribution of costs likely to come from mitigation or adaptation (and acceptable, or 'unavoid-able' impacts) led strategies, this is also a political question. The limits to climate change adaptation when framed in this way are cultural, social and political. This may produce some surprising outcomes. Kuhlicke and Kruse (2009), for example, show how local adaptive actions to reduce flood risk along the Elbe river, Germany, rely mainly on anticipation and assumptions about state support, the latter actually being seen to undermine local resilience. In Australia risk of the re-introduction of mosquitoes carrying dengue disease is increasing as a consequence of government advice for households to adapt to increasing drought risk by installing domestic water tanks, the perfect breeding environment for the *Ae. Aegypti* mosquito (Beebe *et al.*, 2009).

So, how can the limits of adaptation be ascertained? There are big lessons from the past in the failure of sophisticated civilisations from the Greenland Norse to Easter Island and the Maya of Central America. In each case a changing climate interacted with dynamic social pressures to undermine productive systems that overwhelmed adaptive capacity and led towards collapse (Diamond, 2005). Contemporary extractive land management systems are revealing their limits too. Sharer (2006) shows how contemporary industrial agriculture in lowland Guatemala and southern Mexico supports fewer people and generates more local environmental degradation than Mayan farming practices, forcing an inten-sifying drive to fell more forest for short-term productive gain. More than anything these cases tell us that the risk from environmental change is a product of social amplification – the failure to recognise and respond in time to emergent risk – rather than an intrinsic quality of the hazard itself. More contemporary evidence of the limits to adapt alongside climatic variability and extremes comes from the failure of coping and past rounds of adaptation made manifest by natural disasters from regional food security crises, to major hurricanes and floods, or local events such as flash floods, water logging and landslides that are local disasters (ISDR, 2009).

Disasters occur when socio-ecological systems coping capacities are over-whelmed (ISDR, 2004). There are four basic pathways for this failure. First, as a result of a lack of resources – and the marginality that underpins this. This can force people (the poor and marginalised) to knowingly live or work in places exposed to risk in order to access other benefits such as close proximity to livelihood opportunities. Lack of resources also limits people's ability for self-protection. Second, as a result of a lack of information. Proactive adaptation is constrained when new hazards or vulnerability drivers emerge that are not planned for and may not be recognised until it is too late. This was the case in the 2003 heat-wave that claimed more than 35,000 lives in Europe, with earlier events, notably in Chicago, USA (Klinenberg, 2002), failing to stimulate learning and anticipatory adaptation in European cities. At the local level social networks can be as important as formal extension and advisory services for learning. Most acutely information fails when early warning is not provided (IFRC, 2005). Third, as a result of institutional failures. This is the principal reason for physical infrastructural failure – the proximate cause for many events. Institutions fail to enable adaptation when those at risk and managing risk are not able to learn critically, but rather are trapped in cycles of marginal improvements of existing behaviour (see Chapter 4); when those at risk and their advocates cannot hold risk managers to account; and when information and resources cannot be used effectively or equitably (Wisner, 2006). Fourth, as a result of the speed of development and application of appropriate technological innovations. In South Asia, in the space of a generation cell phone technology has enabled mobile phones to spread from being the preserve of the wealthy to a ubiquitous feature of urban and rural life alike with knock-on benefits including providing early warning for disaster risk (Moench, 2007). These accounts indicate the complexity of identifying limits to adaptation and the great sociological and geographical variation to be expected.

The argument presented in this book responds to these four strands of enquiry starting from a perspective of wishing to understand, rather than measure adaptation. This requires a broad lens, close to Fankhauser's comment that adaptation as a research field can be interpreted as a revision of sustainable development. Following the critical literature on sustainable development (for example, Grin *et al.*, 2010), climate change adaptation is seen as a process not an object, with discrete capacities, actions and outcomes offering windows for observation. Elements that are subject to being contested in discourse (as different explanations for events and situation are presented) as well as materially (as different actors compete for the control and use of assets and resources). This approach also builds on a belief that the limits of adaptation are rooted in culture and society; they can be subjective but are mutable (Adger *et al.*, 2009c). The primary aspiration of this work is to open debate on adaptation as a critical process. It uses adaptation as resilience, transition and transformation as a basis for this contribution. In one sense this can be seen as adding a novel line of categorisation to those discussed in the typology above. But a more fundamental aim is to highlight adaptation to climate change as a multi-layered process, with observed acts of adaptation potentially concealing or denying opportunities for alternative pathways that could lead to different social and socio-ecological futures. Making these three levels of adaptation transparent is an initial step in supporting actors at risk and managing risk in questioning the power relations that give shape to adaptation as observed. These are tools for a critical consciousness in climate change adaptation.

Structure of the book

The book is organised into four parts. Part I (Framework and theory) contains Chapters 1 and 2. Chapter 1 seeks to outline the intellectual and policy landscape that has thus far shaped understanding of adaptation to climate change. Following from this, Chapter 2 offers a detailed account of those theories and research agendas in social science that have preceded the current interest in adaptation to climate change, but nonetheless, and often without recognition, continue to shape thinking. Lessons that can be learned from these precursors are then taken into a discussion of the contemporary literature on adaptation to climate change. This chapter ends with the outlining of three broad categories of adaptation based on the intention of the initiating actor: resilience, transition and transformation.

The characteristics and the range of literature from which tools can be built to analyse governance for each type of adaptation are outlined in Table 1.1. Adaptation that enhances resilience is characterised by functional persistence, self-organisation and social learning. Adaptation to promote transition in governance regimes includes self-organisation and social learning but can also benefit from insights provided by literature on governance and socio-technological systems. Understanding adaptive capacity or actions that could result in transformational change in socio-political regimes can usefully incorporate social contract and human security theory in addition to literature on regimes, socio-technological regimes, self-organisation and social learning.

In Part II (The resilience–transition–transformation framework) Chapters 3, 4 and 5 describe the qualities of and develop the resilience–transition–transformation framework; this is explored empirically in Part III (Living with climate change) through Chapters 6, 7 and 8.

Table 1.1 Frameworks of the analysis of adaptation

	Resilience	*Transition*	*Transformation*
Functional persistence	*		
Self-organisation	*	*	*
Social learning	*	*	*
Regime theory		*	*
Socio-technological transitions		*	*
Social contract			*
Human security			*

In this way the book builds its argument across Chapters 3–8, and to some extent its separation under the heading of resilience (Chapter 3), transition (Chapter 4) and transformation (Chapter 5) is heuristic as much as it is analytic. This said there is a logical progression with the analysis of functional persistence, self-organisation in Chapter 3 being added to by the analysis of governance regimes and socio-technological transitions in Chapter 4 and finally socio-contract and human security in Chapter 5.

Each case study is set in a different context: Chapter 6 focuses on the organisation, Chapter 7 on urban settlements and Chapter 8 on the national policy as sites for adaptation. These contexts have been chosen because they highlight the preceding chapter's discussion, but they also serve two other tasks. The first is to indicate the complexity of distinguishing adaptive capacity and action which is always dependent on the viewpoint of the observer. In each chapter elements of resilience, transition and transformation can be found. The second goal is to use these case studies to demonstrate the range of social contexts where adaptation unfolds and analysis is needed. At present the majority of analysis of adaptation focuses on local actions, with the site of analysis being the local community or household.

Part IV (Adapting with climate change) contains the concluding chapter, Chapter 9. This final chapter synthesises the detailed discussion made in each preceding chapter and outlines the research and policy development needs that arise from the central argument that adaptation is a social, cultural and political as well as a technological process.

2 Understanding adaptation

> The adapted man, neither dialoguing nor participating, accommodates to conditions imposed upon him and thereby acquires an authoritarian and uncritical frame of mind.
>
> (Paulo Freire, 1969: 24)

Freire warns us that without a critical awareness, adaptation is hostage to being limited to efforts that promote action to survive better with, rather than seek change to, the social and political structures that shape life chances. Similarly, Clarke (2009: 21) warns that people tend to adapt to poverty by 'suppressing their wants, hopes and aspirations' rather than attempting to change the structures that constrain their life chances. Can the same critique be levelled at adaptation to climate change – that efforts are being directed more towards accommodating risk and its root causes rather than at the root causes themselves? The difference is between responding to drought by proving humanitarian relief to alleviate hunger, and identifying distortions in agricultural trade policy and market conditions that prevent food surpluses from moving to meet human need.

This chapter builds on Chapter 1 by reviewing the academic literature on adaptation and adaptive capacity. The aim is to map out an analytical framework and set of linguistic tools to examine the socio-political nature of adaptation. The framework is developed in Chapters 3–5 and applied in Chapters 6–8. We begin by defining key terms and outlining the broad intellectual legacy that thinking about adaptation can learn from. Contemporary debates are then outlined and the notions of adaptation as resilience, transition and transformation are introduced.

An adaptation lexicon

Adaptation is a deceptively simple concept. Its meaning appears straightforward: it describes a response to a perceived risk or opportunity. The IPCC defines climate change adaptation as 'adjustments in natural or human systems in response to actual or expected climatic stimuli or their effects, which moderates harm or exploits beneficial opportunities' (IPCC, 2008: 869).

Complexity comes with distinguishing different adaptive actors (individuals, communities, economic sectors or nations, for example) and their interactions, exploring why it is that specific assets or values are protected by some or expended by others in taking adaptive actions, and in communicating adaptation within contrasting epistemic communities. It is also important to distinguish between coping and adaptation, and adaptive capacity and adaptive action.

Coping precedes adaptation as a concept in explaining social responses to environmental stress and shock by some 30 years, and continues to be used within disaster studies to describe many of the same processes now captured by adaptation in the climate change community. The latter has to some extent re-invented the wheel in doing this (Schipper and Pelling, 2006). With these two terms in use, approaches have been taken to demarcate separate meanings. For some coping is associated with reversible and adaptation with irreversible changes in behaviour (White *et al.*, 2004). However, work on both adaptation and coping accepts that the transition from reversible to irreversible changes is critical for measuring the collapse of system sustainability (Swift, 1989). In practice, coping and adaptation still exist as parallel concepts serving epistemic communities with different origins but very similar interests and conceptual frameworks (see below). This can be seen by the slowness with which IPCC included the term coping and ISDR the term adaptation in their respective glossaries (ISDR, 2004).

Here adaptation is defined as: the process through which an actor is able to reflect upon and enact change in those practices and underlying institutions that generate root and proximate causes of risk, frame capacity to cope and further rounds of adaptation to climate change. Coping with climate change is defined as: the process through which established practices and underlying institutions are marshalled when confronted by the impacts of climate change.

Adaptation includes both adaptive capacity and adaptive action as sub-categories. Capacity drives scope for action, which in turn can foster or hinder future capacity to act. This is most keenly seen when adaptation requires the selling of productive assets (tools, cattle, property) thus limiting capacity for future adaptive action and recovery.

Adaptive capacity has been conceptualised both as a component of vulner-ability and as its inverse, declining as vulnerability increases (Cutter *et al.*, 2008). This distinction is important in designing methods for the measurement of adaptive capacity and vulnerability, which are generally conceived of as static attributes, and the subsequent targeting of investments to reduce risk. The distinction is less important for theoretical work and methods aimed at revealing vulnerability or adaptive capacity as dynamic qualities of social actors in history. For these projects more important is the recognition that vulnerability and adaptation interact and influence each other over time, shaped by flows of power, information and assets between actors. The relationship between vulnerability and adaptive capacity varies according to size and type of hazard risk and the position of the social unit under analysis within wider socio-ecological systems. Position matters as vulnerability and adaptive capacity at one scale can have profound and sometimes hidden implications for other scales. For example, a family in Barbados may benefit from living in a hurricane-proof house (low micro-vulnerability) but still be impacted by macro-economic losses should tourists be deterred by hurricane risk in an island whose economy specialises in tourism with limited diversity (low macro-adaptive capacity).

The predominant understanding of adaptation is that while it is a distinct concept it is part of the wider notion of vulnerability. The IPCC conceptualises vulnerability as an outcome of susceptibility, exposure and adaptive capacity for any given hazard (and inadvertently compounds the definition through use of the term 'to cope'!):

> Vulnerability is the degree to which a system is susceptible to, and unable to cope with, adverse effects of climate change, including climate variability and extremes. Vulnerability is a function of the character, magnitude, and rate of climate change and variation to which a system is exposed, its sensitivity, and its adaptive capacity. (IPCC, 2008: 883)

Exposure is usually indicated by geographical and temporal proximity to a hazard with susceptibility referring to the propensity for an exposed unit to suffer harm. Adaptation then can either reduce exposure or susceptibility. Adaptive actions to reduce exposure focus on improving ways of containing physical hazard (building sea walls, river embankments, reservoirs and so on); they can also include actions to shield an asset at risk from physical hazard (by seasonal or permanent relocation or strengthening the physical fabric of a building, infrastructure and so on). There is some contention here with individual studies including shielding under exposure or susceptibility. However, this is only problematic when assumptions are not made clear, preventing aggregation of findings (a specific concern for the IPCC which seeks to build knowledge on vulnerability and adaptation from local studies worldwide). Susceptibility can be reduced by a wide range of possible actions including those taken before and after a climate-change-related impact has been felt.

The separation of mitigation and adaptation by the UNFCCC may help international policy formulation, but it is intellectually problematic. Mitigation can most logically be viewed not as a separate domain but as a subset of adaptation. It is an adaptive act aimed at ameliorating or reversing the root causes of the anthropocentric forcing processes behind climate change. Changing lifestyles and technologies to reduce carbon are then acts of adaptation targeted at supporting mitigation. Bridges between adaptation and mitigation are being made. Already the need to consider mitigation when developing adaptation strategy is recognised by the engineering community in the concept of climate proofing (McEvoy *et al.*, 2006). The analysis presented in this book does not include mitigation acts or policy, but does include discussion on the cultural and social norms that shape development worldviews and acts; a line of analysis that is well suited to questions of mitigation and in this way offers an approach that can be developed to connect mitigation to the adaptation agenda.

Adaptation occupies a pivotal position in the coproduction of risk and development (see Figure 2.1). Through this one phenomenon one can gain insight into the social mechanisms leading to the distribution of winners and losers and identify opportunities and barriers to change as both risk and development coevolve. Too often, though, research and policy development on adaptation has focused on narrow technical or managerial concerns; for example, in determining water management or sea-defence and building structural guidelines. Wider questions of political regime form, social values and so on that direct techno-logical and social development and risk have been acknowledged as root causes, but put in the background as too intangible or beyond the scope for adaptation work (Adger *et al.*, 2009c). Within the community of social researchers tackling climate change the social root causes are well acknowledged and amongst many policy-makers they are recognised too. Multilateral development agencies and NGOs such as WRI, WWF, Practical Action and CARE are taking a lead and tools are being developed to help policy-makers, who have understood the complexity of the problem but not had access to the tools to begin planning for adaptation.

The possibility that adaptation can inform political as well as managerial and technical discourses and structures is presented in Figure 2.1 as a distinction between resilience, transition and transformation. Resilience (see Chapter 3) refers to a refinement of actions to improve performance without changing guiding assumptions or the questioning of established routines. This could include the application of resilient building practices or application of new seed varieties. Transition (see Chapter 4) refers to incremental changes made through the assertion of pre-existing unclaimed rights. This might include a citizens' group

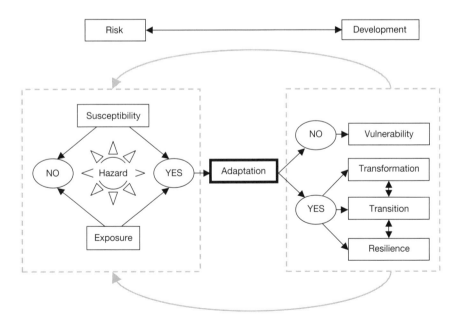

Figure 2.1 Adaptation intervenes in the coproduction of risk and development

claiming rights under existing legislation to lobby against a development that would undermine ecological integrity and local adaptive capacity. Transition implies a reflection on development goals and how problems are framed (priorities, include new aspects, change boundaries of system analysis) and assumptions of how goals can be achieved. Transformation (see Chapter 5) refers to irreversible regime change. It builds on the recognition that paradigms and structural constraints impede widespread and deep social reform; for example, in international trade regimes or the individual values that are constitutive of global and local production and consumption systems. Where adaptation is not undertaken in response to a perceived risk (a hazard event for which a social actor is both exposed and susceptible) vulnerability will remain unchallenged.

The three levels of adaptation are nested and compounding. Nesting allows higher-order change to facilitate lower-order change so that transformative change in a social system could open scope for local transitions and resilience. Compounding reflects the potential for lower-order changes to stimulate or hinder higher-order change. Building resilience can provoke reflection and be upscaled with consequent changes across a management regime, enabling transitional and potentially transformative change – but it could also slow down more profound change as incremental adjustments offset immediate risks while the system itself moves ever closer to a critical threshold for collapse. On the ground mosaics of adaptation are generated from the outcomes of overlapping efforts to build (and resist) resilience, transition, local transformative change and remaining unmet vulnerabilities; mosaics that can change over time as underlying hazards and vulnerabilities as well as adaptive capacity and action change driven by local and top-down pressures.

The discussion of terms so far has been generic with climate change leading potentially to opportunities as well as threats. On the ground opportunities arise for some actors from even the most catastrophic of climate change associated events. This is especially so when accountability and transparency are limited, as they are post-disaster creating gross market distortions: following Hurricane Katrina, a contracting company Shaw charged FEMA US$175 a square foot for temporary roof repairs, material costs were provided by the USA government and workers paid as little as US$2 a square foot (Klein, 2007). Elsewhere increased rainfall and temperatures may extend the growing season, leading to locally increased agricultural productivity, particularly in developed countries that can also capitalise on technological innovation leading to local benefit but further global inequality (UNDP, 2007). There may be more benign opportunities from climate change that need not contribute to greenhouse gas emissions or exaggerate social inequality (such as crop reselection), but these appear trivial compared to present and predicted costs. The annual impact from natural disasters associated with climate change alone accounts for tens of thousands of deaths, millions of people affected and billions of US$ lost, with drought, flooding, temperature shocks and wind storms causing the greatest impacts (Guha-Sapir *et al.*, 2004). By 2080 the number of additional people at risk of hunger could reach 600 million (Hansen, 2007). Climate change threatens the Millennium

Development Goals and most especially the development prospects of a large section of humanity. UNDP (2007) argues that some 40 per cent of the world's population, over 2.6 billion people, will be consigned to a future of reduced opportunity without action to mitigate and adapt to climate change. With this context, our focus on adaptation is primarily as a mechanism to avoid harm.

The impacts of climate change will be felt directly (weather related and sea-level rise events), indirectly (through the knock-on consequences of reduced access to basic needs as critical infrastructure is damaged or employment lost) and as systems perturbations (the local implications of impacts on global commodity prices or international migration). Adaptation therefore needs to insert itself to ameliorate vulnerability caused by each level of impact. However, as one moves from direct through indirect to systems perturbations, climate change impacts interact with other systems features such as development policy, demography and cultural norms. This makes it increasingly hard to identify and communicate the consequences of climate change in isolation so that adaptation becomes both a climate change specific and more generic human process of development. The vastness of climate change and the multitude of pathways through which it can affect life and wellbeing for any individual or organisation make it almost impossible for 'climate change' in a holistic sense to be the target of adaptation. In comparison, international targets for mitigation are relatively simple. Rather, people and agencies tend to adapt to local expressions of climate change – flood events, changing crop yields or disease vector ecologies, often without attributing impacts or adaptation to climate change. This again makes identification, communication and ultimately the development of supporting governance structures for climate change adaptation a challenge unless such efforts are integrated into everyday activities and structures of policy-making.

The antecedents of adaptation

The notion of adapting to environmental stress and shock has been the focus of previous rounds of academic investigation from fields outside climate change. To varying degrees the ideas generated have been recognised and incorporated in the development of the idea within the climate change community. Jeffry and McIntosh (2006) identify relevant literature dispersed across economics (industry sector dynamics, innovation processes and risk-taking behaviour), psychology (characteristics of inventors and risk takers), philosophy of science (roles of innovation/invention), sociology (population dynamics, sociology of groups and networks), anthropology (collapse of complex societies) and evolutionary theory (role of diversity and adaptation in survival).

Despite the rich inheritance of contemporary writing on adaptation to climate change this is rarely explicitly noted. Four streams of thinking on adaptation are examined in this section; the first historical, the others still in use and interacting with the climate change adaptation discourse, but all largely outside the mainstream of writing on climate change adaptation. First to be reviewed here are those perspectives on adaptation that have drawn from the ecological systems (cybernetics and coevolution). This strategy has its roots in early sustainable development theory building with efforts to overcome the false dualism of nature and society. Second is a body of work that uses the language of adaptation and learning to describe policy development over time (adaptive management). Third are those approaches that have come from the interface of international development, governance and disaster studies (coping).

Together these antecedents of the contemporary debate on adaptation in the climate change community make up a conceptual backdrop, one with which to contextualise contemporary literature on adaptation to climate change, and to identify gaps and repetition in the development of the idea and its critiques.

Cybernetics

Academic geography has a long history of engagement with adaptation. In the 1970s and '80s this was first explicitly formulated as part of an experiment with cybernetic theory. Cybernetics drew on evolutionary theory to connect analysis of social and natural systems. It was in part a response to the preceding schools of regional geography and human ecology studies that tended to present the environment as little more than background, assuming its malleability to human intervention. Cybernetics sought to provide a more integrated approach to human–environment relations, and one that could be engaged with in a quantitative manner and so exploit the new computer modelling capacities emerging at that time. Natural disasters, including slow onset drought and food security events, were used to exemplify the need for the more integrated approach offered by cybernetics. Given that the cybernetic approach and contemporary resiliency school (see Chapter 3) have similar roots in ecological theory, the criticisms levelled at cybernetics are especially worthy of consideration.

In 1975, Vayda and McCay first advocated adaptation as a bridging concept between cultural ecology and natural hazards research. They conceptualised the interaction of social and natural systems through backward and forward flows in energy and material. This helped to provide some quantitative modelling purchase but was not further developed. But many elements of adaptation introduced in this period do reoccur in contemporary debates. This includes an interest in the temporal staging of adaptive actions, on the possibility of mal- or sub-optimal adaptation, and in later work on social context as a root cause of adaptive actions. With its base in ecological understandings of systems dynamics this perspective used parsimony (rather than equity) as a measure of effectiveness in adaptation. Under this rubric adaptive actions should not require any unnecessary or excessive commitment of resources. Should initial adaptations prove insufficient additional actions would be taken so that adaptation unfolds in a sequential and rational pattern of increasingly resource-intensive interventions (Slobodkin and Rappaport, 1974). The ecological origins of this approach to adaptation inspired this rationalist logic and also removed any discussion of values or justice. The aim of adaptation was to maintain stasis in the face of environmental perturbations, not to enable progressive change in social or socio-ecological systems. A contemporary critic, Morren (1983), also regarded the cybernetic approach as being limited by focusing on loss reduction not prevention.

Under the cybernetic approach, adaptive capacity was approached through the notion of flexibility: 'uncommitted potentiality for change' (Bateson, 1972: 497). The principal of parsimony meant that loss of flexibility (opportunity for future adaptive actions) was seen as a particularly significant cost of adaptation. Much effort was put into developing typologies of flexibility and adaptation and comparing this with specific environmental pressures. Counter to the rule of parsimony, great variation was observed in the actions taken by people facing similar hazards (Morren, 1983). By supporters of this approach such findings were considered as irrational actions by those at risk. Critics argued that while the cybernetic approach had made progress in providing a framework that recognised social context as a mediating pressure on the environment, shaping risk and adaptation, it did not have the conceptual tools to analyse these relationships. Analysis of adaptation was trapped at the level of information access, transmission and decision-making apparatuses. Deeper social relations of production and power were not included.

One outcome of this failing of the cybernetics approach, which continues to influence work on adaptation and vulnerability to disaster risk today, was to provide the inspiration for the self-styled, alternative school (Hewitt, 1983). The alternative school sought to reveal the structural root causes shaping risk by drawing from neo-Marxist dependency theory. Within this tradition, Watts argued that:

> the forces and social relations of production constitute the unique starting point for human adaptation which is the appropriation and transformation of nature into material means of social reproduction. This process is both social and cultural and it reflects the relationships to and participation in the production process. (Watts, 1983: 242).

For Watts, adaptation went beyond human responses to environmental change or natural hazard to incorporate all processes of environmental transformation and interaction with the natural world including extraction for wealth creation. This key conceptual contribution continues today with the realisation that climate change adaptation is but a part of deeper and broader processes of social change and inertia. In analytical terms the key contribution of the alternative school was to open a theoretical framework grounded in critical theory for the analysis of the structural constraints that they argued determined human capacity and action in response to external environmental shocks. This critical view expanded analysis from the technical attributes that surround specific adaptation decisions, to the social life in which they are embedded. Contributions included critique of the structures of humanitarianism and international development that it was argued allowed vulnerability to persist and did not support progressive adaptation in the face of environmental risk (Susman *et al.*, 1983). This critique has particular salience given the influence of ecological and systems inspired theory on the conceptualisation of adaptation within climate change science today.

Coevolution

Drawing metaphorically from the language of evolutionary biology, coevolution, as proposed by Norgaard (1995), extends the cosmology of adaptation by bringing in values. It also expands the time-horizon and scale of what might be considered adaptive action from the local and immediate to global and long-term interactions. Adaptation in the context of climate change similarly extends co-evolution, by including inanimate natural elements as well as biotic and human ones as subjects and forces for change (Adger and Brooks, 2003). In short, coevolution is found in the reciprocity of interacting components (including human, technological, physical and bio-chemical elements and systems) within evolutionary systems. Norgaard (1995) includes knowledge and values alongside technology, social organisation and the natural environment as categories, sites and drivers for adaptation. Norgaard also moves from a materialist (adaptation can be described through technical changes, for example, in engineering or farming practices) to a relational and constructivist epistemology (where adaptation includes changes in identity and wellbeing including humanity's relation with the non-human) so that:

> a technological innovation or introduction from another region will affect the fitness of various aspects of social organisation, perhaps favoring a different mix of individual and community rights, or favoring more or less hierarchical ways of socially processing information. The changes in social organisation, in turn, might feedback on the fitness of other components in the technological system, or favor some types of values or types of knowledge over others. (Norgaard, 1995: 486)

Adaptation seen through the lens of coevolution is not an end point. It is a transitional and relational episode in history; one that is open to back-sliding, distortion and amplification as outcomes interact with other sub-systems in the coevolving whole. Coevolutionary processes change structure and interaction rules. They typically preclude the possibility of previous system states reoccurring. This is a distinction from the dynamic characteristics of non-evolutionary models where only status can be changed, not guiding rules. The rules in ecological systems are fixed (until our understanding of nature and physics changes) – in social and socio-ecological systems rules of culture and law are mutable.

Coevolution emphasises change. Innovations drive the coevolutionary process, but their drivers (disaster events, macro-economic cycles, household collapse) are often not amenable to planning. This makes it difficult, perhaps impossible, to predict with high confidence what will work best in the other subsystems as adaptations and their consequences coevolve with the whole system potentially never reaching a new equilibrium (Klüver, 2002). This challenge argues for a shift from seeking to predict and control sub-systems, and through this the whole, to a framing that argues for adaptive planning. This is achieved through the maintaining of diversity to keep options open and a preference for monitoring rather than a presumption for managing or resisting changes. Consequently discourse and the flow of information, decision-making capacity and processes and ability to implement decisions are highlighted as subjects for research and policy if adaptation is to be understood and supported.

Norgaard (1995) also reflects on the relationships between humanity, nature and hydrocarbons. Under the coevolutionary epistemology he argues that humanity has coevolved with hydrocarbons, not nature/ecosystems (Norgaard, 1995). In the short–medium term this has been possible because hydrocarbons have shielded (and alienated) us from nature, but the consequences of a failure to select production systems and institutions that coevolve with nature is now being felt (Norgaard, 1995). At this scale of analysis living with climate change includes acts that are not simply adaptive or mitigative, but that underpin generic capacity, such as a movement from material consumption to community as a source of identity including the (re)building of communities of place and personal relationships with nature. Coevolution, then, points to a large gap in contemporary climate change science which has only recently begun to consider the deeper cultural needs of and drivers for adaptation (O'Brien, 2009). It also offers a framing for thinking through this problem.

The abstract nature of coevolution makes for difficult translation into an empirical research framework. While coevolution has been successful at the level of metaphor to frame accounts of adaptive behaviour within complex systems (Pelling, 2003a) and economic-ecological systems interaction at the global scale (Schneider and Londer, 1984) it has more limited applicability as a tool for local analysis. One useful line of analysis highlighted by this lens is the relationship between intention (policies) and emergence (self-organised activity) in policy sectors, the latter in large part accounting for observed divergence from policy during implementation (Sotarauta and Srinivas, 2006) and so revealing tensions between the actions and values of competing adaptive strategies or other behaviour. Jeffrey and McIntosh (2006), in a review of the coevolution of land use and water management, argue that 'noise' from the range of interconnections in any system makes it difficult to distinguish coevolution from state-based dynamic change, and at a more general level they ask what it is that coevolution brings that has not already been proposed through complex systems theory. More contestable is Costanza's (2003) criticism that this approach offers little potential as a planning or predictive tool. To be sure, coevolutionary approaches are more able to capture backward than forward looking assessments of adaptation, but methods have been developed, in particular integrated scenario assessments, that allow some purchase for forward looking analysis of the interaction between sub-systems and constraints on adaptation (Lorenzoni *et al.*, 2000).

Adaptive management

Like cybernetics and coevolution, adaptive management draws from systems theory and recognises the interdependence of the social and ecological. Its focus is also on large and complex socio-ecological systems dynamics; for example, watershed or forestry management. Its major contribution is in taking us from abstract, modelling or conceptual work to that based firmly in the empirical reality of decision-makers who wish to mainstream adaptation into changing socio-ecological contexts.

First developed in the late 1970s to support decision-making under uncertainty for natural resource management (Holling, 1978), adaptive management is part of a wider body of literature on organisational management that sees social/organisational learning as a key attribute for systems survival (Argyris and Schön 1978) (see Chapters 3 and 6). This is often explained as the spread of successful innovations from individuals to become common practice; for example, where a new agricultural or management practice is copied until it becomes the norm. Under adaptive management individual and organisational learning is both encouraged from planned actions (such as change in the regulatory environment) and in response to unplanned environmental surprises (natural or technological disasters). While not specifically formulated with climate change in mind, the aim of providing a conceptual framework and subsequent management guidance for decision-making in contexts where information is scarce and contexts are dynamic is analogous to the challenge facing forward looking climate change adaptation (Pelling *et al.*, 2007b).

Learning is enabled in adaptive management through ongoing policy experiment. This usually takes the form of centrally developed management innovations that are piloted locally. If successful they may be replicated or up-scaled across the management regime. Underlying hypotheses explaining relationships between management actions and environmental systems are in this way compared and adapted to over time. This should produce continuous and anticipatory adaptation (Kay, 1997); indeed as the environment changes in response to social adaptations this would demarcate a coevolutionary system over time.

A range of interpretations of the adaptive management approach exist. Learning is framed as an activity at the interface of environmental and economic policy, through to wider questions of democratic principles, scientific analysis and education (Medema *et al.*, 2008). Walters and Hilborn (1978) distinguish between different degrees of formality in learning, between passive and active adaptive management, with active approaches using formal scientific methods to evaluate experiments and, it is claimed, providing more reliable information for decision-makers. Medema *et al.* (2008) describe active approaches as experience–knowledge–action cycles. In all cases high levels of stakeholder involvement are required for the surfacing of hypotheses and the translation of experimental findings into policy learning.

Evidence from existing experiments in adaptive management offer an early opportunity to observe the challenges likely to present themselves if adaptation to climate change were to become mainstreamed into development. Some very significant challenges to adaptive management have been identified by Walters (1997), Lee (1993) and Medema *et al.* (2008).

In a review of 25 adaptive management regimes in riparian and coastal ecosystems of the USA, Walters (1997) found only two that were well planned with programmes being distracted by focusing on the process of model development and refinement rather than field testing and application. Walters argues that failure in the take-up of adaptive management by senior decision-makers is caused by a combination of the perceived short-term expense and risk of undertaking experiments, concern that the acknowledgement of uncertainty and acceptance of experimentation inherent in adaptive management may undermine management credibility, and lack of participation from stakeholders. Lee (1993) also analyses the barriers to take-up and adds that the high costs of information gathering and monitoring and associated difficulties in acquiring funding have also inhibited the implementation of adaptive management approaches. Medema *et al.* (2008) summarise these challenges into four barriers for implementation of adaptive management, each with an associated research agenda. These are presented in Table 2.1. Their most important call is for long-term research on the outcomes and challenges of adaptive management which unfold slowly and very differently in individual contexts; a proposal that fits well with the need to shift from indicating adaptation capacity to verifying the outcomes of adaptive actions.

The institutional and economic constraints identified in Table 2.1 are all amenable to policy that can support experimentation and make learning from error an acceptable method for living with change. Where climate-change-associated uncertainty is increasing, the efficiency argument may also move in favour of a more adaptive management approach.

Adaptive management also helps to provide insight into a key element of adaptation to climate change – multi-stakeholder collaboration for social learning. Evidence suggests that many of the challenges to this aspect of adaptive management are common to other development approaches that seek to incorporate or be led by community actors. Such challenges are most well studied in international development contexts (for example, Mungai *et al.*, 2004) and often revolve around the distribution of power between local and management actors worked out through the division of labour and responsibilities, and control of information and decision-making rights (Pelling *et al.*, 2007b). In a study of seven community-based forestry management organisations supported as part of adaptive management programmes in the western USA, Fernandez-Gimenez *et al.* (2008) found that the best outcomes measured by benefits in social learning, trust and community building, and application and communication of results came from projects where local actors had been given an opportunity to participate, not only in data collection and monitoring but also in design and objective setting, and where projects were supported by commensurately large budgets. Of those projects with much more limited financial support the best results were found where community members participated in multiple roles.

Table 2.1 Barriers for the implementation of adaptive management

Challenge	Barrier for adaptive management	Research agenda
Institutional	Rigid institutions (cultural values and more formal rules). Lack of stakeholder commitment to share information over the long term.	What institutional arrangements are best suited to implementing adaptive management?
Evidence of success	The use of 'soft' conceptual and qualitative modelling makes it difficult to communicate outcomes. The boundaries between adaptive management and background processes can be difficult to distinguish.	Methodologies are needed to gather evidence for and communicate the outcomes of adaptive management to stakeholders.
Ambiguity of definition	Multiple, ambiguous definitions make it difficult for resource managers to understand how they can apply this approach.	Is ambiguity a potential strength indicating diversity? Refining the typology of approaches associating themselves with this adaptive management will help add clarity.
Complexity, costs and risk	Experimentation can be ecologically and economically risky. Adaptive management is slow and planning costs are high compared to centralised management.	An honest dialogue is needed on the appropriateness of concepts from complexity science such as sub-optimality, uncertainty and diversity.

(Source: based on Medema *et al.*, 2008)

From this more bottom-up perspective the key challenges for adaptive management – and by implication for integrating adaptation into development planning more generally – can be identified:

- the need for higher level organisations to be receptive to local viewpoints and undertake learning in response,
- the challenges of maintaining local engagement over extended time-spans, and
- determining and securing the needed level of technical assistance and science capacity to ensure the validity and credibility of community-led efforts.

Fernandez-Gimenez *et al.* (2008) also point to the opportunities that adaptation can open. They note that community-led approaches to adaptive management can be a source of local skill training and employment generation in the establishment of an ecological monitoring workforce. These could in part offset or help to justify the financial costs of adaptation in development.

Coping mechanisms

The notion of coping has acquired a sizable and well developed literature. It describes the strategies used by those living with rapid onset disasters such as flash floods, and chronic disasters, including drought and food insecurity (Wisner *et al.*, 2004). This matches well the dual interests of adaptation to climate extremes and base-line change. Coping has also been used to explore social change in relation to wider impacts of social violence and personal tragedy (Lee *et al.*, 2009). Despite this wealth of knowledge of direct relevance to climate change adaptation, learning has been limited (Schipper and Pelling, 2006). This makes it important to identify what, if any, are the similarities between coping and adaptation, and what adaptation could usefully take from this literature; and also to make clear the boundaries between these two concepts.

Within the natural disasters and food security literature numerous models for coping have been proposed since the 1970s. These have variously been framed by entitlements (Sen, 1981), human ecology (Hewitt, 1983), game theory (Uphoff, 1993) and livelihoods analysis (Leach *et al.*, 1997). Across these theoretical realms models tend to be agency focused, the majority operating at the household level and to differentiate coping either by stage or sector of action. Burton *et al.* (1993) is one of the most encompassing models, connecting slow cultural change with rapid adjustments. This four-stage model commences with loss absorption where hazard impacts are tolerated, absorbed as part of the ongoing coevolution of socio-ecological systems with no tangible impacts or observed, instrumental adjustments. Stage two, loss acceptance, is reached once the negative effects of a hazard are socially perceived but losses are borne without active mediation. The third stage of loss reduction commences once losses are perceived to be higher than costs for mitigation; this is the focus of most disaster reduction work. A final stage of radical change is reached once hazard impacts can no longer be mitigated and major socio-economic changes are experienced either through impact or attempts to minimise disaster loss. This broad view of coping is useful in identifying coping as simultaneously a long-term (cultural) and short-term (economic) process of realignment to changing environmental conditions. This model also flags the importance of perception on action. The implication of a temporal dimension opens the possibility of tipping points where one stage flips into another through changes in vulnerability or hazard.

Alternative categorisations of coping offer typologies of action; for example, Wisner *et al.* (2004) identify four kinds of coping action: disaster prevention and loss management (for example, hazard mitigation schemes, early warning systems), diversification of production (for example, the promotion of mixed cropping, livelihood diversification), development of social support networks (for example, informal reciprocity or state welfare) and post-disaster actions to contain loss (for example, opportunistic livelihoods, insurance, novel social organisation). This approach has the advantage of providing technical detail but is restricted to Burton *et al.*'s stage of loss reduction and possibly radical change. While these models are designed to accommodate action at multiple spatial scales they less easily reveal the trade-offs and interactions of coping interacting across scale. Livelihoods models are one response to this challenge and explicitly situate agents (normally households) within an institutional context. Coping responses are located at the interface of actors and institutions (Leach *et al.*, 1997).

While a successful concept, coping is ultimately misleading as a metaphor for social responses to environmental change at it implies that actors are getting by, doing okay. This can be the case, with agriculturalists, for example, deploying coping mechanisms to get through the low-productivity periods in the annual agricultural cycle (Davies, 1993). But often, acts labelled as coping require the expenditure or conversion of valuable assets to achieve lower-order outcomes, undermining current capacities and future development options. This ratchet effect (Chambers, 1989) is socially amplified when multiple individuals, households or businesses deploy similar economic strategies – selling assets or changing livelihoods – and so undermining market value. Competition can turn into collaboration with virtuous magnifier effects through the use of social capital, which can be built up and whose impact can be extended through multiple simultaneous actions. There are, however, limits even to individual and societal stocks of social capital so that continuing environmental stress or repeat shocks can lead to a cascade of failure as social and economic assets are expended. Figure 2.2 indicates a sequence of coping acts that can lead to collapse as assets are depleted in the face of unrelieved stress.

Swift (1989) argues that household collapse becomes inevitable once core social and economic assets are lost and is observed even when macro-economic conditions improve, revealing how individual vulnerability, or capacity to cope, operates with a degree of independence from structural conditions. Households, especially poor households, live with many kinds of risk as well as a desire to fulfil unmet needs and wants. So it is that households have to play off expenditures on immediate household maintenance against investment to recover lost resources or offset anticipated risk, and this can make it more difficult to replace savings or productive assets once they have been expended through coping. The potential for social capital to be undermined through ever more destructive rounds of coping links household collapse to that of collectively held assets such as social cohesion or notions of community. Commencing with a shift in investment and use from bridging to bonding capital that amplifies cultural difference and competitive group behaviour (Goodhand *et al.*, 2000), subsequent coping detracts from more fundamental aspects of local social capital through a withdrawal of investment in short-term (health) and long-term (education) social capital, and finally in fragmentation of the most basic social unit – the household. As with the economic cascade, cultural contexts will determine the order movement. For

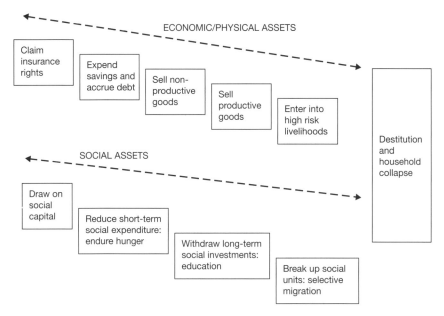

Figure 2.2 The coping cascade: coping and erosion of household sustainability
(Source: based on Pelling, 2009)

example, child sharing is a well-developed coping mechanism in the Caribbean that need not signify approaching household collapse. Here the extended family, not the household, is the basic unit of social organisation (Pelling, 2003b). Broadly, though, as a household approaches collapse subsequent acts are more difficult to reverse.

The delicate balance between the terms coping and climate change adaptation (see Table 2.2), and the negotiation of the intellectual division of labour between them can be found in some early writing on adaptation. Kelly and Adger (2000) define coping capacity as the ability of a unit to respond to an occurrence of harm and to avoid its potential impacts, and adaptive capacity as the ability of a unit to gradually transform its structure, functioning or organisation to survive under hazards threatening its existence. This distinction builds on earlier work; for example, working on food security, Gore (1992, in Davies, 1993) offers a distinction based on the actor–institution relationships. Coping is the means to survive within the prevailing systems of rules; adaptation is indicated when institutions (cultural norms, laws, routine behaviour) and livelihoods change. This distinction is becoming increasingly accepted. Under this rubric an example of coping might be selling cattle during drought, with adaptation signified by migration or a change in livelihood to supplement or replace dependence on livestock. Critics of this division argue that, on the ground, the distinction between coping and adaptation in terms of the depth of consequence for actors is

Table 2.2 Distinctions between coping and adaptation

Coping	Adaptation	Source
The ability of a unit to respond to an occurrence of harm and to avoid its potential impacts	The ability of a unit to gradually transform its structure, functioning or organisation to survive under hazards threatening its existence	Kelly and Adger (2000)
The means to survive within the prevailing systems of rules	Change to the institutions (cultural norms, laws, routine behaviour) embodied in livelihoods	Gore (1992)
The range of actions available to respond to the perceived climate change risks in any given policy context	Change to the set of available inputs that determine coping capacity	Yohe and Tol (2002)
The process through which established practices and underlying institutions are marshalled when confronted by the impacts of climate change	The process through which an actor is able to reflect upon and enact change in those practices and underlying institutions that generate root and proximate causes of risk, frame capacity to cope and further rounds of adaptation to climate change	Pelling (2010)

greatly influenced by the viewpoint of the observer. This blurs the practical utility of the empirical boundaries between coping and adaptation, producing a potential lack of analytical and policy clarity (for example, Saldaña-Zorrilla, 2008).

Yohe and Tol (2002) offer a nuance on the distinction between coping and adaptation described above. They see adaptive capacity as describing the set of available inputs that determine coping capacity which itself is manifest in the range of actions available to responding to perceived climate change risks in any given policy context. Adaptive capacity is determined by underlying social factors: resources, institutions, social capital, human capital, risk spreading, information management and awareness. Their availability is context specific and path dependent. Coping capacity is defined by the range of practical measures that can be taken to reduce risk. The range, feasibility and efficiency of these measures is determined by adaptive capacity. This logic reveals some insightful outcomes in the relationships between inputs and actions (adaptation and coping). Enhanced investment in the 'weakest link' component of adaptive capacity has the advantage of raising coping capacity across the board – or at least until the next weakest link emerges to limit coping. By the same token investing in one component in isolation need not increase coping capacity. Adding to the resource base may, for example, have no effect on coping capacity if institutional processes or decision-making structures block implementation.

The distinction being made by these authors reflects other attempts to disentangle distinct relationships between actors and their environment. This helps provide some depth to the more narrowly focused challenge of coping/adaptation in climate change. The interest of Freire (1969) was to make transparent the potential role of education in society – much like the climate change problem, his concern was to see development as a process that contained what the poor knew and what they imagined they could do with knowledge. The distinction between 'adapted man' (that is, someone who has learnt to live with the current system) and 'critical consciousness' has parallels with coping and adaptation. Adapted man corresponds with coping – where successive rounds of coping, that is, of accommodating one's life to live with hazard, describe well the ratchet effect undermining assets and human wellbeing. Critical consciousness – the ability to see one's position in society as a function of social structures as a prerequisite to seeking ways of making change in those structures – has great parallels with the institutional dimensions of adaptation described above. The difference is that climate change has to date been driven predominantly by a concern for maintaining efficiency in the output of economic systems and livelihoods rather than in the balance of power between actors or as embodied in institutions. Thus the current modes of defining adaptation go only halfway to meet Freire; they acknowledge the action to change institutions but do not emphasise the potential for emancipation that this could bring – nor indeed that this could be a parallel and even motivating goal for climate change adaptation.

The systems worldview that has had a great influence on recent thinking about human responses to climate change also recognises the potential for more profound change (for example, Flood and Romm, 1996; Pelling *et al.*, 2007b). Argyris and Schön (1996) identified three kinds of learning, termed first, second and third loop learning. Only the first two are encompassed routinely in the distinction between coping and adaptation in climate change literature. First loop learning corresponds with coping – learning to improve what you already do. Second loop learning corresponds with adaptation – learning to change the mechanisms used to meet your goals. Third loop learning – learning that results from a change in the underlying values that determine goals and actions – is less clearly expressed within current adaptation theory.

The lack of emphasis in climate change literature on adaptation as critical consciousness or third loop learning is likely a reflection on the difficulty of making clear empirical associations between climate change related impacts and social change of this order. Chapters 5 and 8 aim to provide one step forward in opening this discussion. There is also the possibility that the climate change community – which has its eyes tightly focused on the IPCC process, and which in turn is a product of negotiated content between science and governments – has not found analysis of power as part of adaptation to be a priority. It risks alienating the political and technical decision-makers for whom the IPCC endeavour is designed to support.

Another area where coping is still a predominant term, and one where further development could prove insightful for work on climate change, is the psychological literature. This work views coping as an interior action determined by the interaction of cognitive and emotional process, but acknowledging interaction with socialised values, access to information and social–historical context. Individual ability to cope with stress associated with catastrophe has been described as psychological resilience (Walsh, 2002). This literature is most developed in the USA, with Hurricane Katrina stimulating many studies including Lee *et al.* (2009) who identified psychological resilience as an outcome of survivors' perseverance, ability to work through emerging difficulties and ability to maintain an optimistic view of recovery. Amongst this group those who suffered human loss were least able to cope, with property loss having only a minor impact on capacity for psychological recovery. Other hurricane events in the US have shown that survivors who reported more resource loss also reported higher levels of active and risk-reducing behaviour (Benight *et al.*, 1999). This has important implications for the appropriateness of mainstream methodologies for measuring disaster impact and for disaster response and recovery efforts which predominantly focus on economic and physical rather than social and psychological aspects.

Psychology has begun to offer some insight into the factors leading to individual wellbeing and empowerment post-disaster, although the link to material coping actions is as yet less well defined. Psychological traits associated with coping following Hurricane Katrina included a heightened sense of control over one's destiny and of personal growth. These in turn were attributed to survivors who were problem-focused, accepting of loss, optimistic and held a religious worldview (Linley and Joseph, 2004). In the general population talking, staying informed and praying enabled coping, emerging as predictors of decreased psychological stress during post-disaster relocation (Spence *et al.*, 2007), with spirituality particularly significant for older African American Katrina evacuees (Lawson and Thomas, 2007). In a comparison of psychological resilience pre- and post-Katrina, Kessler *et al.* (2006) found reduced thoughts of suicide after the disaster in survivors expressing faith in their ability to rebuild their life and a realisation of inner strength. This is important in providing an empirical link for adaptation, between internal processes of belief, identity and self-worth and external actions, in this sad case illustrated through suicide rates. Outside the US, following the 2003 earthquake in Guatemala, feelings of self-control and self-assurance were also found associated with adaptation outcomes of 'successful survivors' who reconceptualised the crises as opportunities for acquiring new skills (Vazquez *et al.*, 2005). This work provides one approach for promoting a progressive response to climate change through acknowledging the interplay of social and psychological root causes (Moos, 2002), but this has yet to be systematically applied (Zamani *et al.*, 2006). It provides an initial evidence base to begin a characterisation of specific psychological orientations associated with adaptation and linking interior and exterior expressions of adaptation, taking us closer to gaining some leverage on the ways in which individuals and social collectives might move between different cognitive, emotional and potentially intellectual states; the latter opening scope for the study of shifts between 'adaptive man' and critical consciousness or first, second and third loop learning.

In order to incorporate deeper levels of change while retaining close links to the existing literature adaptation to climate change is defined here as: The process through which an actor is able to reflect upon and enact change in root and proximate causes of risk.

This formulation sees coping as the range of actions currently being enacted in response to a specific hazard context. These are made possible by existing coping capacity (which may extend beyond the range of coping acts observed at any one time). Adaptation describes the process of reflection and potentially of material change in the structures, values and behaviours that constrain coping capacity and its translation into action. Coping then is an expression of past rounds of adaptation. Both adaptation and coping will unfold simultaneously and continuously in shaping human–environment relations, they will interact and on the ground they may be hard to separate as reflection and application occur hand-in-hand. Still, from an analytical perspective and for policy formulation there is a value in distinguishing these two components of human–environment relations.

The coproduction of vulnerability/security by coping and adaptation brings the possibility that adapting to climate change can undermine as well as strengthen capacities and actions directed at coping with contemporary climate related risks. Coping may be limited for longer-term gain or a result of ignorance or injustice in the implementation of adaptation. This can be seen in the loss of income accepted by low-income families who are able to provide an education for their children. This is an adaptive action that constrains contemporary coping capacity, but with the aim of providing future gains that will provide the means for better family wellbeing including capacity to cope with uncertainty and shocks associated with the climate change. More likely, the immediacy of political life will produce a tendency for coping that distracts from or undermines the critical reflection and long-term view of adaptation. The danger is that coping is felt to be sufficient so that the potentially difficult questions and changes in development that adaptation might bring are temporarily evaded. At the scale of large social systems, this tension is illustrated by the trade-off between short-term social disruption and the long-term easing of socio-ecological friction proposed by Handmer and Dovers (1996) (see below).

Adaptation as a contemporary development concern

The preceding discussion on the antecedents of adaptation reveals the framing behind contemporary understandings of adaptation. This is not always explicitly acknowledged in the climate change literature but can be felt, for example, in the pervasive influence of systems thinking. Systems theory has had a far-reaching influence with its promise of providing a mechanism to integrate the social and natural. It is used in cybernetics, adaptive management and to a lesser extent in coevolution as well as in contemporary adaptation studies, particularly through work on resilience (Folke, 2006). The aspect of adaptation given prominence in each application reflects fashions in social scientific research as much as the underlying use of systems theory in each case. Cybernetics, developed at a time when positivism was seen as providing new scope for generalisable theory, sought to apply a value neutral, technical epistemology. It is reductive, opening scope for mathematical modelling of behaviour but not able to incorporate the significance of competing values and power asymmetries in shaping action. Adaptive management acknowledges the role of difference in access to information and decision-making capacity in shaping adaptive processes and outcomes, but does not have power as a focus of analysis; like cybernetics the focus is on technical aspects but in this case with a view to informing policy learning. Coevolution orients adaptation less towards the search for ways in which to manage risk and change and is more interested in adaptation as a process, a state of living with uncertainty. It stands back from technical and management analysis to examine the bigger picture of historical change where contesting values are included as a driver for change alongside knowledge, technology, organisational forms and the natural environment. Coping is the outlier in offering a legacy for adaptation that is grounded not in systems theory but in development studies. Connections between nature and society are context specific and hard to generalise from, although a common language has been developed through work on vulnerability (partly originated as a critique of the cybernetic school) that acknowledges both the roots of coping in political-economy but also the influence of values and social viewpoint in shaping decisions and options for adaptation. These four approaches highlight a tension in understandings of adaptation which persists today. This is between policy friendly but reductive analysis on the one hand, and holistic, value sensitive and critical but potentially unwieldy work on the other.

The antecedents also offer guidance on the qualities that promote adaptive capacity. These include parsimony (that the best adaptive choice is that which expends least resource); flexibility; diversity; monitoring to facilitate appropriate change (as distinct from managing to maintain stasis); learning as a facet of policy systems and organisations as well as individuals; and a realisation that observed adaptation, while a positive attribute, is also a sign of stress and a play-off that can signify approaching collapse and reduced wellbeing. These ideas have been taken up by resilience thinking and have a strong influence on contemporary framings of adaptation (see below). They also set adaptation apart from other logics for assessing development, perhaps most important that of economic maximisation, a cornerstone of economic globalisation. This argues economies should invest in what they do best, leading to a concentration of assets and closing off options for diversity and flexibility in the productive sectors (Pelling and Uitto, 2001).

The aim of this section is to examine the contemporary conceptualisation of adaptation in detail. We review a typology of adaptation, discuss the influence of resilience on the conceptualisation of adaptation and the significance of social thresholds as tipping points for adaptive change, and compare economic and ethical frameworks for evaluating adaptive choices. This sets the context for the proposal of the three adaptation pathways – resilience, transition and transformation – that are then developed throughout the remaining chapters.

A typology of adaptation

Following the technocentric bias of its antecedents, much of the early work on adaptation was theorised as a technical act of adjusting economic or other functions to a changing external environment. This bias has gradually been eroded. An important literature in this regard has been that focusing on adaptation in developing country contexts (Adger *et al.*, 2003; Nelson *et al.*, 2007) including urban (Satterthwaite *et al.*, 2009) and rural (Tanner and Mitchell, 2008) contexts. Contributions have also been made from work demonstrating the need to include values, feelings and emotions in decision-making (O'Brien, 2009).

As summarised in Chapter 1, a sizable and fundamental literature on adaptation is directed towards differentiating adaptations (see Smit *et al.*, 2000; Smit and Pilifosova, 2001). Table 2.3 presents a typology of adaptation to be taken forward in this framework, and also distinguishes between the impacts of different adaptive actions. These include actions that respond to perceived positive as well as negative impacts of climate change; those that are felt directly (heat events), indirectly (the price of food or water) or through perturbations in socio-ecological systems (political instability). They are acts unfolding within many sectors (urban planning, water management, agriculture development, transport planning and so on) and using a range of vehicles (technical innovation, legislative reform, market adjustment, professional training, behavioural change). They describe both the nature of an adaptive action and its scope of impact.

Table 2.3 A typology of adaptation

Criteria	Options
Nature of Adaptive Action	
Degree of collaboration	individual or collective
Degree of focus	purposeful or incidental
Degree of forethought	spontaneous or planned
Phasing	proactive or reactive
Scope of Impact	
Target	proximate, intermediary or root causes
Timescale	of risk
Future wellbeing	immediate or delayed
Social consequences	climate-proofing or maladaptation
Developmental orientation	regressive or progressive
	autonomous or integrated

Adaptation is purposeful when directed towards a recognised hazard or opportunity (retro-fitting of a building) and incidental when undertaken in response to some other pressure that has consequences for exposure, susceptibility or adaptive capacity (economic opportunities leading to migration out of a flood-prone location). Proactive adaptation is that which takes place before a risk manifests into hazard (disaster risk reduction); reactive adaptation takes place during or after an event (disaster reconstruction). The scope of adaptive action can be distinguished between that which seeks to change material assets or practices set against less direct institutional change (see Pelling and High, 2005). This is reflected in the potential targets of climate change which may be proximate (crop variety), intermediary (local decision-making systems) or root causes (political–economic structures and development visions). Timescale acknowledges that adaptation can have immediate (changing built forms) or delayed (investing in health and education) benefits. The impacts of adaptation on the future wellbeing of others are indicated by acts that could be termed as climate proofing (the integration of mitigation) or maladaptation (adaptation that increases vulnerability); socially regressive or progressive depending on redistributive consequences, and autonomous to (isolated and contained) or integrated in (undertaken with awareness of and aiming at synergies with the actions of others) development.

Resilience and adaptation

Resilience is popularly understood as the degree of elasticity in a system, its ability to rebound or bounce back after experiencing some stress or shock. It is indicated by the degree of flexibility and persistence of particular functions. That resilience is not simply synonymous with adaptation has been well demonstrated by Walker *et al.* (2006a) who argue that adaptation can undermine resilience when adaptation in one location or sector undermines resilience elsewhere, where management focus on a known risk distracts attention from emergent hazard and vulnerability, and that increased efficiency in adaptation (through risk management, for example) can lead to institutional or infrastructural inertia and loss of resilient flexibility.

Resilience has been contrasted both with stability and vulnerability. Stability, according to Holling (1973), is an attribute of systems that return to a state of equilibrium after a disturbance. This compares with resilient systems that might be quite unstable and undergo ongoing fluctuation but still persist. Stability is more desirable in circumstances where environmental perturbations are mild; resilience is most useful as an attribute of systems living with extremes of impact and unpredictability. Within the disaster risk community, resilience has been interpreted as the opposite of vulnerability. The more resilient, the less vulnerable. But this belies the complexity of the conceptual relationship between these terms which have also been constructed as nested – with vulnerability being shaped by resilience (Manyena, 2006) which for some in turn incorporates adaptive capacity (Gallopin, 2006). Stability and vulnerability provide useful bounding concepts for resilience. They suggest that resilience is about the potential for flexibility to reduce vulnerability and allow specific functions to persist. What it does not tell us is how these functions are identified or who decides (Lebel *et al.*, 2006). This requires a more critical engagement with social processes shaping resilience (see Chapters 3, 6, 7 and 8).

Working with the idea of resilience, and especially efforts that seek to measure it are made difficult because of its multifaceted character. The processes and pressures determining resilience for a unit of assessment change with spatial, temporal and social scale – a community may be resilient to climate change associated hurricane risk (through early warning and evacuation, for example) but less resilient to the long-term inflections of climate change with the local and global economy. The subjects of analysis are also wide, bringing diversity but also fragmentation to the study of resilience. Cutter *et al.* (2008) identify studies attributing resilience and related metrics to ecological systems (biodiversity), social systems (social networks), economic systems (wealth generation), institutional systems (participation), infrastructure systems (design standards) and community competence (risk perception) (Folke, 2006; Paton and Johnston, 2006; Rose, 2004; Perrow, 1999; Vale and Campanella, 2005).

One of the first critical engagements with resilience from the perspective of environmental risk management came from Handmer and Dovers' (1996) proposal of a three-way classification of resilience. This insightful framework has echoes of Burton *et al.*'s (1993) classification for coping and still offers a great deal. It highlights both the contested and context specific character of adaptation that this book argues for, and is worth describing in some detail. The three-way classification presented resilience as: (1) resistance and maintenance; (2) change at the margins; and (3) openness and adaptability.

Resistance and maintenance is commonplace, particularly within authoritarian political contexts where access to information is controlled. It is characterised by resistance to change; actors may deny a risk exists with resources being invested to maintain the status quo and support existing authorities in power. When risk is undeniable these systems typically delay action through a call for greater scientific research before action is possible. Vulnerability can be held at bay by resource expenditure; for example, in food aid or through containing local hazard risk through hard engineering 'solutions'. But this can generate additional risks for other places and times through global flows of energy, resources and waste. This type of resilience offers an easy path for risk management, there is little threat to the status quo and considerable stress could be absorbed. However, when overcome the system would be threatened with almost complete collapse – Diamond's (2005) thesis on the collapse of ancient civilisations reminds us of this possibility.

Change at the margins is perhaps the most common response to environmental threat. Risk is acknowledged and adaptations undertaken, but limited to those that do not threaten core attributes of the dominant system. They respond to symptoms, not root causes. Advocates argue that this form of resilience offers an incremental reform, but it is as or more likely to delay more major reforms by offering a false sense of security. Preference for near-term stability over radical reform for the wellbeing of future generations provides a strong incentive for this form of resilience. This approach is well illustrated by the Hyogo Framework for Action on Disaster Risk Management, which sets forth an international agenda agreed by nations for managing disaster risks including those associated with climatic extremes. Not surprisingly given the vested interests of dominant voices in the international community for the status quo, the framework is limited. It calls for the integration of risk management policy into development frameworks, the increasing of local capacity for risk reduction and response, and for new systems of disaster risk identification and information management (ISDR, 2005).

Social systems displaying openness and adaptability tackle the root causes of risk, are flexible and prepared to change direction rather than resist change in the face of uncertainty. That this mode of resilience is so rare is testament to the huge inertia the results from personal and collective investment in the status quo. Large fixed capital investments make change difficult as do investments in soft infrastructure – preferences for certain types of education or cultural values making shifts painful in industrial societies. Dangers also lie with this form of resilience: instability will lead to some ineffective decisions and maladaptation would need to be prepared for within individual sectors as a cost of wider systems flexibility. These are both worries that decision-makers have cited in making it difficult for them to commit to adaptive management strategies, as described above.

Handmer and Dovers prefigure their account by a caution that while the three classifications are designed to cover the full range of policy responses to the adaptation challenge, most actors will operate in only a small part of this range. This points to a central dilemma for progressive adaptation – that the comfort zone for adaptive action is relatively small because both those with power and the marginalised are wary of the instability they fear from significant social change (see Chapter 5). Resilience then has the possibility of both identifying the scope for flexibility within the socially accepted bounds of stability but also making transparent for all social observers the range of choices foregone. Mapping the characteristics of social systems that are more or less amenable to these three forms of resilience is a key foundation for the analytical framework development in this book which places emphasis on the processes through which systems undertake or resist adaptive change.

More contemporary work on resilience and its relationships with vulnerability and adaptation have also applied critical reasoning. This has focused on the advantages of inclusive governance. This, it is argued, facilitates better flexibility and provides additional benefit from the decentralisation of power. On the down side, greater participation can lead to loose institutional arrangements that may be captured and distorted by existing vested interests (Adger *et al.*, 2005b; Plummer and Armitage, 2007). Still, the balance of argument (and existing centrality of institutional arrangements) calls for a greater emphasis to be placed on the inclusion of local and lay voices and of diverse stakeholders in shaping agendas for resilience through adaptation and adaptive management (Nelson *et al.*, 2007). This is needed both to raise the political and policy profile of our current sustainability crisis and to search for fair and legitimate responses. Greater inclusiveness in decision-making can help to add richness and value to governance systems in contrast to the current dominant approaches which tend to emphasise management control. When inevitable failures occur and disasters materialise this approach risks the undermining of legitimacy and public engagement in collective efforts to change practices and reduce risk. This takes us back to Handmer and Dovers' (1996) analysis of the problem of resilience and shows just how little distance has been travelled in the intervening years.

Adaptation thresholds

Acts of adaptation are stimulated by the crossing of risk, hazard and/or vulnerability thresholds. Each threshold is socially constructed, a product of intervening properties including identification, information and communication systems, political and cultural context and the relative, perceived importance of other risks, hazards and vulnerabilities that compete for attention. The existence of social thresholds explains the 'lumpiness' of human experience, where history does not unfold as a gradual story but in fits and starts. Forward looking adaptation, or the impacts of climate change resulting from a lack of sufficient adaptation, may be catalysts for the breaching of thresholds.

Risk is ever present in society. The level of risk that is accepted by different social actors determines the first threshold (see discussion on coping) and is shaped by whose values and visions for the future count in society (Adger *et al.*, 2009a). For any social group the level of acceptable risk can change as scientific innovation, media interest and public education influence awareness amongst the public and decision-makers. Communication between science, decision-makers, the media and the public is determined by norms of trust. Trust is built over time by the everyday performance of scientific or government bodies but is easy to lose (Slovic, 1999). Where there is a confidence gap in advisory bodies, the government or science, popular regard for new risk announcements will be greeted with scepticism (Kasperson *et al.*, 2005), with a preference for self-reliance or fatalism amongst those at risk and potentially resistance to any coordinated adaptation. It is here that dedicated intermediary organisations or individuals that can translate climate science into the language of target audiences (such as agricultural extension agencies) play a significant role in shaping people's willingness to reduce risk (Huq, 2008). Indeed part of the challenge facing adaptation to climate change is the need to communicate without confusing, and the science community that has championed climate change research thus far has not found this easy (Hulme, 2009).

Climate change is felt locally through many environmental indicators. Figure 2.3 represents how just one – say precipitation – is influenced by climate change and how this is related to the timing and scope of coping and adaptation. In this case climate change produces reduced hazardousness at minimum extremes (drought) but increased hazardousness at maximum values (flood). In this way new hazard thresholds challenge existing hazard management strategies which are breached until the changing hazard threshold is recognised (E1) and responded to (E2). The distance between these two points reflects the risk acceptance and communication thresholds described above. A final threshold that determines adaptation comes from changing vulnerability profiles.

Demographic and economic change in particular influence the likelihood of adaptation. This is often not integrated into accounts of adaptive capacity and action (see Figure 2.3) but is particularly important in rapidly changing contexts such as rapid urbanisation, economic restructuring or where social tensions might lead to armed violence. The vulnerability threshold suggests there is a critical mass of assets or people at risk and of risk management capacity that are needed for adaptation to be likely. This also has consequences for the kind of adaptation undertaken. Thus a small coastal settlement may undertake independent, spontaneous adaptations to protect livelihoods in the face of sea level rise, but should this area be subject to investment by the corporate tourism sector and subsequent high levels of labour in-migration adaptation may become more coordinated, collective and planned.

Figure 2.3, although stylised, is useful in demonstrating several other attributes of adaptation (Füssel, 2007). It shows the disproportionate ability of extreme over average climatic conditions to stimulate adaptation, the need to consider natural climatic variability and anthropocentric climate change together in planning adaptations, and the continuous process of review and response needed of adaptation to climate change as hazard thresholds change (E3). The fuzzyness inherent

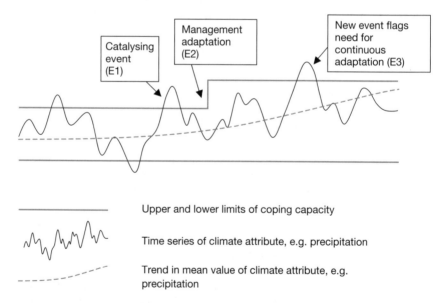

Figure 2.3 Adaptation thresholds
(Source: based on Füssel, 2007)

in labelling adaptation as reactive or proactive is revealed with the decision to adapt being both a reaction to the preceding extreme event and a proactive anticipation of future risk. A reactive motivation can lead to a proactive adaptation. This said, the time needed to make decisions to adapt to climate change and complete adaptive measures such as major infrastructure works or the reform of housing stock is often several years if not decades so that incremental adaptation may be dangerous and costly (Reeder *et al.*, 2009). In contrast, planning over extended timeframes opens decision-making to uncertainty. As the limits of scientific knowledge are reached so decisions are based increasingly on value judgements. These in turn are shaped by the structures and norms of governance systems and cultural–historical expressions of acceptable risk that inform and legitimate adaptation (Paavola and Adger, 2006). Ultimately this directs scrutiny to questions about who it is that determines the principles upon which adaptive choices are made as much as the nature of the decisions themselves.

Evaluating adaptive choices: economics and ethics

There are two bases for evaluating between adaptive choices: economic costs and human rights. At the global scale both approaches have already been used to argue for mitigation (Stern, 2006). Lack of agreement on global responsibilities for the distribution of costs of adaptation (which have not been fully calculated but likely far outweigh those of mitigation) mean the case for adaptation has been less forcefully argued using either approach, although human rights has been used to frame accounts of climate change impacts as unjust, for example by the UN Human Rights Commission in its resolution 7/23 (UN Human Rights Commission, 2009).

At the regional level and within countries there is some experience in the use of cost–benefit analysis (CBA) as a tool for adaptation decision-making (Splash, 2007). CBA tries to establish the costs of alternative adaptive measures and how much damage can be averted by increasing the adaptation effort given a specific climate change scenario. CBA works for individual sectors where costs and benefits can be derived from market prices; it is harder when multiple sectors are included and when market prices are unavailable – for example, in placing a value on human health or wellbeing – and where the items being compared are incommensurable (Adger *et al.*, 2009c). Despite such limitations, some sophisticated methods are emerging which can at least show clearly what is known and provide a logical framework for political judgement. For example, it has been suggested that the range of choices for adapting to heat stress in the UK (though not their social and environmental costs, including potential for maladaptation) is likely to be maximised in future global contexts characterised by active free markets and entrepreneurialism, but more limited if strong environmental regulation becomes the norm (Boyd and Hunt, 2006). CBA has also been used effectively to argue for proactive adaptation through investment in disaster risk reduction as an alternative to managing disaster risk through emergency response and reconstruction. The World Bank and US Geological Survey calculate that an investment in risk management of US$40 billion could have prevented US$280 billion in losses during the 1990s alone, a CBA ratio of 7:1. In high risk locations advantages of proactive risk reduction are even higher, Oxfam calculates that construction of flood shelters costing US$4,300 saved as much as US$75,000 a ration of 17:1 (DFID, 2004a). These are compelling ratios but do not allow estimation of costs for specific investments before disaster strikes and in this respect their weight in decision-making is limited.

Given the methodological constraints on economic assessment for the costs and benefits of adaptation options can ethics help? Caney (2006) argues that people have a moral right not to suffer from the adverse effects of climate change. However, a central dilemma for investing in adaptation based on human rights when resources are scare is whose rights to prioritise. What is the basis on which to decide? Is it fairer to target interventions to reduce risk of climate change impacts and aid adaptation amongst the most vulnerable (as Rawls would argue), or aim to generate the maximum collective good (following the utilitarian philosophy of Bentham). The latter approach may well target those who are only marginally vulnerable. It is justified by the assumption that the overall increase in wellbeing would provide a resource for compensating those negatively impacted by this decision. The utilitarian approach is one origin of economic cost–benefit analysis.

There are many strands to systematic thinking on justice that could inform decision-making for adaptation. The dominance of OECD countries in inter-national policy and the academic literature positions the Western philosophical tradition closer to the existing intellectual core, and the relative potency of justice arguments thus framed. This is not to deny that non-Western philosophies, many perhaps not formalised, will shape local decisions and actions. Indeed their interaction with top-down policy based on Western ideas of justice may be a source of tension or misunderstandings. There are also inspiring and profound differences that can inform questions of sustainability and adaptation from non-Western sources. For example, the Buddhist aim to decrease suffering (including unmet desires) through individual control of the birth of desires (Kolm, 1996) presents a radical departure from dominant Western logics which aim to address perceived need not through individual self-knowledge, chosen restraint and a revelation of happiness, but through the social rights of access, distribu-tion and procedure; or worse through imposed coping and restraint in the worst forms of adaptation. Meeting these Western elements of justice has further been constrained by a framing of the solution in dominant liberal political-economies that assumes needs must be met through increasing material wealth and energy consumption – an error identified by many Green philosophers and lying at the heart of Norgaard's (1995) observation of the lack of sustainability and risk produced by humanity's dangerous coevolution with hydrocarbons.

Returning to the question of how to prioritise resources to support adaptation, a review of Western philosophical traditions suggests there is no simple or single answer. Justice theories distinguish between logics of equality, priority, sufficiency and desert. Egalitarian principles demand that justice be concerned with equality of some relevant distributable elements. Prioritarian principles claim the importance of supporting adaptation for the least advantaged subjects. Sufficientism holds that every subject must have a sufficient, yet not equal, share of support in adaptation. The justness of a society depends on its capacity to give people the support they deserve (Grasso, 2008). Theories are further differentiated by feminist and communitarian arguments that justice is contextual (Konow, 2003) and over the nature of equality. With respect to egalitarian principles, Sen (1987) differentiates between equality in outcomes (equal post-adaptation vulnerability), the meeting of needs (some basic level of security for all) and command over resources (equality in adaptive capacity). Individual principles can be reinforcing strengthening arguments. For example, prioritarian logic is supported by Shue's 'guaranteed minimum' principle of equity (Shue, 1999) which, from a sufficientarian standpoint, states that those who have less than enough for a decent human life be given enough. This general principle of justice has been applied to climate change adaptation to support the argument that interventions prioritise the most socially vulnerable first (Paavola and Adger, 2006; Paavola *et al.*, 2006, Adger *et al.*, 2009c).

Of the approaches outlined above, it is worth spending some more time with Rawls who helps add clarity to the different realms within which justice for climate change adaptation is manifest. Rawls argues that for any social system justice requires both the application of distributional and procedural justice. Rawls made these two elements of justice the cornerstones of his *Theory of Justice* (1971). Procedural justice talks to the institutions and behaviours that frame decision-making, distributional justice talks to the outcomes of these decisions. Under Rawls, a just society is one where procedural justice is embodied in an egalitarian social contract based on reciprocity, so that individual or sectional interests are given les weight than the overriding drive for distributive justice (Chapters 5 and 8 develop the importance of the social contract for establishing justice in adaptation). This understanding of procedural justice places with individual citizens the responsibility for producing specific declinations of equality and defining the basic structures for their society. With this responsibility comes the right to craft and argue for alternative development and adaptation visions. The climate change literature highlights three aspects of procedural justice that it is argued determine the quality of procedural justice (Paavola, 2005; Paavola *et al.*, 2006):

- Recognition demands acceptance of minority perspectives in planning and decision-making processes, implying that the views and aspirations of the most marginalised and vulnerable be acknowledged.
- Participation requires access to knowledge so that all affected parties can formulate informed viewpoints and be involved in the decision-making process with engagement ranging from consultation to local autonomy.
- Distribution relates to whom holds and uses power to ensure equal participation and recognition of the weakest in decision-making.

This triad can be applied across scales from global negotiations on adaptation regimes to local planning for adaptation in development and together with distributional justice is necessary to underpin legitimacy and popular consent for international, national and local adaptation strategies (Adger *et al.*, 2006).

As with economic analysis, ethics does not provide an easy answer but rather a logic around which options can be discussed with more transparency. Experience from the disaster risk reduction community suggests that while ethical arguments may be useful in the shaping of priorities, once political attention is gained economic based arguments are more persuasive in advocacy.

Three visions of adaptation: resilience, transition and transformation

Adaptation offers a unique lens for understanding and influencing development, and operates at different levels of engagement with specific social systems. Table 2.4 identifies three levels at which adaptation can intervene in development – through enabling resilience, transition or transformation. These three levels are introduced below to provide a framework for assessing adaptation aims and outcomes and then developed in the following chapters. No level of adaptation is intrinsically more desirable than the others; everything depends on context and viewpoint. Very little in social life is uncontested, so it is unlikely there will be many cases where there is an easy consensus on which form of adaptation is required. Indeed different actors may be working to build capacity and action for adaptation at different levels simultaneously; for example, when local community actors organise to challenge local power asymmetries as part of an agenda for transformative adaptation in a locale which is also the target of government sponsored technical reforms to livelihood or infrastructure provision seeking to build resilience (and possibly mollify local acts of transformation).

Adaptation to build resilience acts at the most contained level, seeking only change that can allow existing functions and practices to persist and in this way not questioning underlying assumptions or power asymmetries in society. Transformation is the deepest form of adaptation indicated by reform in over-arching political-economy regimes and associated cultural discourses on development, security and risk. Transition acts at an intermediary level of engagement, focusing on the governance regime but through acts that seek to assert full rights and responsibilities rather than make changes in the regime. In asserting rights or undertaking responsibilities that might previously have been neglected or disallowed incremental transformation is a possibility. Each form of adaptation can include changes to values, institutions, behaviour and assets so that it is the scope and range, rather than depth of change that distinguishes each adaptive form.

While it is possible to distinguish individual ideal types theoretically and empirically, for a specific policy domain or social group different levels of adaptation may not be clearly bounded and can influence one another. Transformative adaptation will at a minimum include a critical reflection on existing institutions

Table 2.4 Attributes of adaptation for resilience, transition and transformation

	Resilience	*Transition*	*Transformation*
Goal	Functional persistence in a changing environment	Realise full potential through the exercise of rights within the established regime	Reconfigure the structures of development
Scope	Change in technology, management practice and organisation	Change in practices of governance to secure procedural justice; this can in turn lead to incremental change in the governance system	Change overarching political-economy regime
Policy focus	Resilient building practice Use of new seed varieties	Implementation of legal responsibilities by private and public sector actors and exercise of legal rights by citizens	New political discourses redefine the basis for distributing security and opportunity in society and social-ecological relationships
Dominant analytical perspectives	Socio-ecological systems and adaptive management	Governance and regime analysis	Discourse, ethics and political-economy

and practices working at the levels of transition and resilience. Over time, resilient and transitional adaptations may highlight wider challenges, build capacities and weaken barriers for reform and so feed the adaptive transformation of regimes. It is also possible that apparent success at one level of adaptation may hide problems at other levels so that resilience can inhibit transition or transformation. The power of resilience to suppress deeper changes in the institutions and values that shape development and risk management is reinforced by its attractiveness as a solution to climate change risks for donors and government precisely because it does not challenge the wider status quo. The technical and organisational innovations required by resilient adaptation are less politically challenging, often more visible and quicker to implement than transitional and transformative adaptations.

Part II

The resilience–transition–transformation framework

3 Adaptation as resilience

Social learning and self-organisation

> The ability of a social or ecological system to absorb disturbances while retaining the same basic structure and ways of functioning, the capacity for self-organization, and the capacity to adapt to stress and change.
>
> (IPCC, 2008: 880)

The IPCC definition of resilience, presented above, is forward looking, placing emphasis on capacities rather than outcomes of self-organisation and social learning. Within this, adaptation is positioned as a sub-set of resilience (along with functional persistence and self-organisation). Following from this definition, the framework suggested uses the idea of resilience to capture the first kind of adaptation to be discussed in detail in this book. In our use, adaptation as resilience is a form that seeks to secure the continuation of desired systems functions into the future in the face of changing context, through enabling alteration in institutions and organisational form.

Elsewhere (Olsson *et al.*, 2006; Nelson *et al.*, 2007) the need to recognise adaptation as including more fundamental shifts has led authors to include the areas of transition (Chapter 4) and transformation (Chapter 5) as sub-sets of resilience. These are not problematic arguments, but the framework presented in this book finds the distinctions so central to the nature of adaptation that separate identities are proposed for these three forms of adaptation. This conviction comes from empirical work where imposing resilience in the face of great social inequality is very problematic (see Chapters 7 and 8).

The IPCC definition, and ours, both point at the influence of socio-ecological systems (SES) theory on the understanding of resilience. The three cornerstones of the SES construction of resilience are included: functional persistence, self-organisation and adaptation (if seen as an outcome of social learning) (Folke, 2006). The contribution of SES theory to understanding resilience will be reviewed here and also in following chapters where the elements of resilience described in SES theory contribute to understanding transitional and transformative adaptation. The defining quality of resilience that distinguishes it from transition and transformation is a desire to maintain functional integrity.

This chapter begins by presenting a vision of adaptation as resilience. The contribution of SES theory to this construction of resilience is then examined with a detailed assessment of social learning and self-organisation. This framework is then combined with organisational management theory to build a framework for examining adaptation as resilience.

A vision of adaptation as resilience

Resilience seeks to protect those activities perceived by an actor to be beneficial for human wellbeing and ecological sustainability but threatened by contemporary or future pressures associated with climate change. The vision of adaptation as resilience is to support the continuation of desired systems functions into the future through enabling changes in social organisation and the application of technology. Such changes are facilitated through social learning and self-organisation (see below) to enable technological evolution, new information exchange or decision-making procedures. More than this, and within the limits of bounded systems, such as development policy for a single watershed or a dairy farming business, achieving resilience may require change in values and institutions within managing organisations, and this can include the challenging of established priorities and power relations and potentially lead to a redistribution of goods and bads (Eakin and Wehbe, 2009). In this way, adaptation as resilience has the potential to contribute to incremental progressive change in distributive and procedural justice within organisational structures. When individual cases that build resilience through internal value shifts are upscaled through government action or replicated horizontally, real opportunities can open for contributing to transitional or transformative change in society (see Chapters 4 and 5), though outcomes can be regressive as well as progressive for sustainable development.

Adaptation as resilience can also allow unsustainable or socially unjust practices to persist (Jerneck and Olsson, 2008). This is perhaps easiest to understand in social contexts where entrenched power asymmetries and exploitative economies are manipulated by the elite to maintain power, even when this undermines sustainability. Such outcomes are less likely when local or national decision-making is held to account, but resilience can still undermine long-term sustainability while appearing to meet the demands of adapting to climate change. This can happen when sustainability challenges are recognised but the transactions costs (including political costs) of change are perceived to be higher than doing nothing, with the least bad option being to adapt within available constraints until perceived thresholds of sustainability are breached, forcing change. For example, in the use of desalination plants to compensate for water demand, the proximate need is met but at a cost of high energy use and pollution of the marine environment. The dynamism of climate change and the unpredictability of local impacts provide the additional rationale of uncertainty to justify resilience as the preferred form of adaptation.

The SES science base that has come to influence thinking about resilience in the climate change literature (Gunderson and Holling, 2002) is closely connected to the adaptive management literature outlined in Chapter 2. SES offers a rich and elegant theoretical landscape and one that continues to expand (Liu *et al.*, 2007). Some have pushed resilience theory towards a recognition of transitional adaptation (for example, Olsson *et al.*, 2006) but in this chapter we focus on SES resilience theory contributions to understanding how valued functions can be helped to persist. SES theory emphasises that ecological and social systems are inextricably linked and that their long-term health is dependent upon change, including periods of growth, collapse and reorganisation (Walker *et al.*, 2006b). In addition to space and time, sociological conceptions of scale also consider how humans symbolise and make sense of reality at different organisational levels (Pritchard and Sanderson, 2002; Cumming *et al.* 2006).

Both a strength and weakness of SES is its presentation as an apparently value neutral, realist epistemology, a product of its origins in systems theory. This has produced a rational and structured framework for understanding human action, one that is particularly attractive to climate change research in offering an approach for integrating human and environmental elements into quantitative modelling of futures scenarios under climate change (Jannssen *et al.*, 2006). A parallel literature that has more recently been brought into an understanding of resilience is that from organisational theory which shares a realist and apparently value neutral epistemology, but is otherwise a much looser body of work sometimes reflecting individual views without being explicitly grounded in a philosophical tradition of enquiry. Organisational theory is reviewed at the end of this chapter, and both literatures are combined in Chapter 6 to analyse the production of adaptive capacity within two contrasting organisational forms.

In thinking through a framework for examining adaptation as resilience built from SES and organisational management theory two limitations inherent in the epistemologies of both approaches must be considered. First, while power is acknowledged, in particular by SES, both literatures are infused with a sense of technical optimism that can downplay the contested character of social life and socio-nature relations. The messiness of decision-making (O'Brien, 2009) is not easily captured. Apparent value neutrality in both cases conspires with technical optimism to emphasise technological innovation and efficiency over critical analysis that might place more weight on the political-economy and cultural root causes of risk and its perception. In this way SES theory has been criticised for a weak integration of social science theory and a tendency to allow for an over-simplification of complex social phenomena (Harrison, 2003; Jannssen *et al.*, 2006). Second, and related, both approaches focus on relational social space but limit analysis to the outer world of interactions between individuals, groups and institutions. Inner worlds of emotion and affect – value, identity, desire, fear – that give shape or meaning to, as well as being drivers for, public actions including adaptation choices (Grothmann and Patt, 2005) are difficult to include.

Framing of resilience

Thinking on resilience within climate change has been influenced by two schools: disaster risk and SES. Disaster risk itself includes varied interpretations of resilience including as a capacity for absorbing disturbances and shocks (Birkmann, 2006) and as the opposite of vulnerability, capturing all those acts and capacities that seek to reduce vulnerability to risk (Adger *et al.*, 2005c). More recently both disaster risk and climate change have been influenced by SES theory so that an additional reading of resilience in the face of natural disasters and climate change has become associated with systems regenerative abilities and capacity to maintain desired functions in the face of shocks and stress (Birkmann, 2006), the meaning used here. In this way SES has acted as a bridge between climate change adaptation and disaster risk theory (and with wider literature on natural resource management). Both interpret adaptation as a process as well as a product of social relations and as a dynamic property such that adaptive capacity can change over time in response to shifting risks and capacities (Pelling, 2003b; Young *et al.*, 2006). Arguably another commonality is a failure to question the framing values and political context of decision-making and fall short of addressing adaptation as transformation (Manuel-Navarrete *et al.*, 2009).

Adaptive capacity then is best indicated not by goodness of fit to current or predicted future threats but by flexibility in the face of unexpected as well as predicted hazards, vulnerabilities and their impacts (Janssen *et al.*, 2007). This opens questions about the trade-offs to be made between flexibility, adaptation and welfare (Nelson *et al.*, 2007). Walker *et al.* (2006a) argue that adaptation can undermine net resiliency by shifting resources and so decreasing capacity or increasing risk in another place or sector, and through over-adaptation and lock-in such that a system becomes unable to adapt to novel threats. For example, in southeastern Australia rounds of engineering based solutions have been used by government to respond to a rising water table and salination. This has created a state of lock-in, making it increasingly difficult for the management system to conceive or invest in a non-engineering response. A highly adapted but fragile system is the result – one that is vulnerable to collapse through dependent co-evolution (Anderies *et al.*, 2006), an example of Handmer and Dovers' (1996) account of resilience as resistance and maintenance.

SES theory on resilience applies thresholds to describe movement from one systems state to another (see Chapter 2). This helps theorise what it is that leads one system to respond to the local impacts of climate change risk through resilience and another through transition or even transformation. Empirical work shows that identifying the location of thresholds before change is difficult because of the multiple and non-linear feedback mechanisms active within SES, so that the ways discourse, institutions and practical action interact are not always transparent or predictable (Nelson *et al.*, 2007). However, evidence does indicate that to activate adaptive capacity requires a social or environmental trigger (a change in attitudes, policy, market conditions or environmental risk and impact) and the appropriate institutional framework.

Nelson *et al.* (2007) contrast deliberate and inadvertent crossing of thresholds from resilience into transition. They argue that deliberate crossing is an indication of both greater adaptive capacity and higher levels of resilience. Two case studies are compared to reach this conclusion. Deliberate transition from agriculture to tourism is exemplified through the actions taken by a local authority in Arizona, USA, in changing its development strategy and support from local agriculture to tourism base. Inadvertent transition is noted in the abandonment of an agricultural economy in Jordan precipitated by unsustainable resource use. The Arizona case shows an actor overcoming the inertia inherent in an established system to move into a more advantageous economic position. No clear point of movement is identified, however, to mark the change from resilience to transitional adaptation, although it is suggested that while both resilience and transitional adaptation rely on the same kind of adaptive capacities it is social systems with greater intensity of vertical organisation (such as a functioning system for information exchange and participation in development planning from local to regional and national levels of government) that are more likely to be able to cross thresholds into transitional adaptation.

Two elements of SES resilience theory that deserve closer attention are social learning and self-organisation. These ideas have been paralleled in other literatures – for example, self-organisation in social movements, participatory and communicative planning (Pugh and Potter, 2003) – and much of the emphasis on trust and relationships that underlies social learning echoes work on social capital which has also been applied to adaptation (Adger, 2003; Pelling and High, 2005). To this extent these ideas represent widely accepted social phenomena key to the understanding of any collective dynamic. They are at work within transitional and transformative as well as resilient adaptations; the distinction between these levels being the subject and context rather than the object of analysis.

Social learning

Social learning is a property of social collectives. It describes the capacity and processes through which new values, ideas and practices are disseminated, popularised and become dominant in society or a sub-set such as an organisation or local community. The outcomes of rounds of social learning are the common values, beliefs and behavioural norms that shape the institutional architecture of social life (Wenger, 1999). Social learning is also ascribed to the socialised process of learning and associated change. This is clearly seen in differences over scale where local worldviews or value systems fit uneasily within dominant discourses of development or culture (Argyris and Schön, 1996). Such diversity can be a resource when alternative behaviour is well suited to meet the challenge of changing environments, but also a compounding factor in institutional inertia and potential barrier to the flexibility needed for resilience (Olsson *et al.*, 2004). At the heart of the contribution of social learning to studies of adaptation lies a tension between dominant and alternative or novel ways of seeing and being, and the potential this opens for individual social actors to shape the trajectory and content of collective learning (challenges for the use of social learning as an analytical tool are discussed in the case studies presented in Chapter 6).

Much of the literature on social learning is interested in improving the efficiency of established practices rather than seeking new practices to resolve underlying sustainability challenges, and in this way it meets the goals of resilience (Armitage *et al.*, 2008). A smaller literature examines the role of social learning in enabling transitional and transformational adaptation. According to Diduck *et al.* (2005) such changes that focus on reform of institutions and organisational frameworks are characterised by:

* high levels of trust,
* willingness to take risks in order to extend learning opportunities,
* transparency required to test and challenge embedded values,
* active engagement with civil society, and
* high citizen participation.

Transformative adaptation that builds on alternative values connects individual to social learning – personal beliefs to culture. In thinking through the relationship between learning and political change, Freire (2000 [1969]) calls for critical reasoning. This, Freire argues, is not the default orientation for problem-solving held by the marginalised or powerful – both prefer to make adjustments within the confines of established norms and structures. Freire terms this kind of problem-solving adaptive ingenuity – finding new ways to fit within and gain advantage from dominant structures without challenging them. Thus critical reasoning is a necessary factor in transformational adaptation but will be absent or marginalised into silence in resilience. There are significant obstacles to be overcome in promoting critical reasoning. The powerful as well as marginalised and vulnerable can be frightened by the uncertainty of change, and change itself can be captured by vested interests. To offset this Freire calls for transformation to come from a dialogue between the marginal and powerful and also between ideas and practice (Freire, 1970). Notions of transformation that have shaped participatory development and specific tools including citizen's fora, citizen's budgets and deliberative decision-making and polycentric governance (Bicknell *et al.*, 2009).

Responding to the novel hazards of climate change requires social learning systems that can respond to the multiple scale and sectors through which risk is felt and adaptations undertaken. Not least to address the challenge of integrating local community level and scientific knowledge and balance strategic thinking with local needs so that decisions are taken at an appropriate level in the organisational hierarchy (Cash and Moser, 2000). Bringing together and making use of local and scientific knowledges is not easy. It is difficult for individuals and organisations to handle both kinds of knowledge and this is exacerbated by inbuilt power imbalances that tend to give greater weight to science over local knowledge (Kristjanson *et al.*, 2009). In response, Wenger (2000) has called for organisations, individuals or tools that can work across this epistemic divide – so called boundary objects.

Self-organisation

Self-organisation refers to the propensity for social collectives to form without direction from the state or other higher-level actors. This can include new canonical (formal) organisational forms such as registered community development groups or trade associations, and shadow (informal) organisations such as networks of friends and neighbours that work independently to or cross-cut canonical organisation. Most research on organisations and adaptation focuses on canonical forms which are visible and easy to access, exemplified by the literature on adaptive management outlined in Chapter 2. However, the generation of novel ideas or practices that are in conflict with or undervalued by canonical organisation often first emerges from the unmanaged space of shadow organisations. Shadow systems are supportive of innovation because they are typically rich in trust, cut across canonical organisational structures and are hidden from formal oversight, allowing experimentation and risk-taking with novel ideas and practices (Shaw, 1997). Successful experiments in shadow systems may in turn become coopted and formalised within the canonical system. This can provide opportunities for the replication of adaptations, but through formalisation of individual roles and relational commitments will limit flexibility and change the social relations which led to the original innovation and potentially undermine long-term sustainability. Alternatively, shadow systems can remain marginalised and informal, operating in parallel with canonical systems. This is especially so under transitional and transformative adaptation where emergent forms are a site for the challenging of established discursive and material power (Pelling *et al.*, 2007).

Self-organisation can evolve slowly in response to changing social values and organisational forms driven by demographic shifts or changes in popular ideology, but also more rapidly. This latter opportunity has been observed following disaster events when new forms of social organisation emerge as dominant forms fail (Pelling and Dill, 2009). Emergent organisation ranges from spontaneous solidarity as neighbours undertake first response, to coordinated networks of NGOs and state agencies in recovery (see Chapters 5 and 8). Capacity to self-organise, like social capital, is particularly difficult to measure in society for this reason – much capacity is hidden and latent, its emergence dependent upon wider social and political context and the nature of threats and opportunities presented to society. It is not possible to measure capacity for self-organisation from existing organisational forms alone. Berkes (2007) notes that because social capital can remain latent in society, social relations that might have been used in the past can be reinvigorated as new threats or needs arise. This was the case in Trinidad and Tobago when networks originally established to deal with coral reef management in Trinidad and Tobago have also played a key role in disaster preparedness (Adger *et al.*, 2005b). Existing organisational forms can also serve to hinder the emergence of novel self-organisation through institutional inertia so that observed high levels of organisation may not alone indicate high levels of capacity for self-organisation to respond to future climate-change-related pressures.

Social learning and self-organisation reinforce each other so that a social system exhibiting rich capacity for social learning is also likely to have considerable scope for self-organisation. The extent to which social learning can be fixed through self-organisation is tracked through three elements of capacity to change: consciousness, institutionalisation and implementation. Consciousness is the capacity to reflect on the outcomes of and alternatives to established norms and practices and sets the limits for subsequent alternative visions or discourses. Bateson (1972) described this as dutero-learning – making learning to learn an act of adaptation. There is no normative assumption on the scope or depth of learning so that this can include adaptive ingenuity and critical consciousness. Institutionalisation is the capacity to move from recognition of constraints to affect change in the institutional architecture that frames implementation through the reproduction of existing, or insertion of new, values and practices. All three processes can unfold within the canonical and shadow systems. Their interaction provides reinforcing or contradictory realms for experimentation and learning and for novel values and practices to emerge (Pelling *et al.*, 2007).

Organisations as sites for adaptation

Organisations operate at scales from the household to firms and national and international bodies. They are often seen as agents in the construction of adaptation for subsidiary actors – for example, by enforcing environmental management regimes or regulating land markets – but in this and the subsequent chapters we also focus on the ways in which internal social relations shape information flow, agency and the direction an organisation can take in adapting itself to a changing external environment. An important distinction is to be made between organisations and institutions.

Following North (1990), institutions are defined as the rules of the game (formal and informal) that influence adaptive behaviour. Organisations are the collective units, embodying institutions, that are vehicles for adaptation. Organisations are not monolithic; they contain potentially competing agents and interests so that internal adaptive change brings new risks as well as opportunities which are not experienced evenly within organisations even when the stated focus of change is on the external environment. Internal differentiation also means that adaptations undertaken by individuals are not always replicated throughout the organisation with consequences for efficiency as well as equity in adaptation. This is the case for households, firms and public or civil society organisations alike.

If organisations and individuals or social groups within them can learn, how might learning be observed? At a surface level, learning is observed through changes in behaviour (signifying implementation and assuming consciousness and institutionalisation). In addition to this behaviouralist Gross (1996) focuses on externally validated, physical behaviour, and following Maturana and Varela (1992) and Ison *et al.* (2000), internal actions are also interpreted as learning. That is, we can learn in relation to different modes of interacting with the world: emotional and conceptual as well as physical. Our learning corresponds to differences in the way that we act (consciously or unconsciously) within these modes, which in turn arise in response to our ongoing experience. The judgement of what constitutes behaviour lies with the observer in question, but the definition does not rule out internal and tacit activities such as conscious or unconscious cognition, emotional affect or the formation and operation of personal relationships, for example. Identifying different realms of behaviour is important in sharpening our focus on the site(s) where adaptation can be observed; not only in material actions, but in contrasting attitudes or views that have not been allowed translation into action. In this way Pred and Watts (1992) identify the behaviour of marginalised actors who need to keep low visibility in the face of surveillance by more powerful actors, and the potential importance of private language as a mechanism for resistance that could form a potential resource for adaptation when organisational relationships or external contexts change.

Constructing the learner as an individual or social entity links individual learning to social processes of change that emerge at the collective level. Thus social adaptation can be seen as collective learning. This is not a claim that individual and collective behaviour are qualitatively the same, but recognises the interaction of learning and adaptive behaviour at these different levels. In this way, adaptation to climate change and variability can be read at different levels of learning operating as a range of system-hierarchic scales – the behaviours of components and subsystems of the system, as well as changes to the emergent properties of the system – and this can be used to unpack different adaptive trajectories: international, national, local. It may be that adaptive behaviour emerging at one scale – say the local – is the result of learning that has been ongoing amongst a range of actors networked across a range of scales. Additionally, adaptation at one spatial (or temporal) scale can impose externalities or constrain adaptive capacity at other scales. In short, the system-hierarchic scale where adaptation is or is not enacted is a socio-political construction (Adger *et al.*, 2005a).

Organisations are spaces of engagement where learning and adaptive capacity can be constrained as well as enhanced (Tompkins *et al.*, 2002). A useful distinction is between organisations and communities of relationships acting within or across them in supporting, antagonistic or ambivalent ways. Communities describe those collectives through which close relationships reinforce shared values and practices; although reinforcement may not necessarily contribute to the organisation's (or even the community's) adaptive capacity, it can lead to closed thinking and the suppression of questioning established norms – a property referred to as groupthink by Janis (1989). In a less formal context a similar phenomenon is described by Abrahamson *et al.* (2009) who observed how closed social networks amongst the elderly led to the self-reinforcing of myths of personal security regarding vulnerability to heatwaves in the UK as new information or ideas were treated with caution.

For Wenger, learning in a community arises through participation and reification, the dual modes through which meaning is socially negotiated. Participation refers to 'the process of taking part and also to the relations with others that reflect this process. It suggests both action and connection' (Wenger, 2000: 55). Participation is thus an active social process, referring to the mutual engagement of actors in social communities, and the recognition of the self in the other.

Reification is the process by which 'we project our meanings into the world, and then we perceive them as existing in the world, as having a reality of their own' (Ibid., 58). Thus reification can refer to the social construction of intangible concepts as well as the meanings that members of a community of practice see embedded in physical objects.

Communities of practice are often not officially recognised by the organisations they permeate (Brown and Duguid, 1991). Their official invisibility in the shadow system can be thought of as being made up of constellations of communities of practice held together by bridging ties of social capital. The link between communities of practice, informal networks and unofficial activity in organisational settings is an important association to make in tracing the workings of the shadow system in building adaptive capacity. Wenger (2000) uses the language of social capital to help define the characteristics of individual communities of practice, which, he argues, can be defined by a shared identity and held together by bonding capital. The influence of personality traits and the role of personal and professional sources of trust in bridging across communities within the public sector are discussed by Williams (2002). It is the quality, quantity and aims of individuals connected together in communities of practice, and their linking of boundary people and objects, that determine the influence of the shadow system on adaptive capacity.

Pathways for organisational adaptation

Figure 3.1 summarises the preceding discussion by identifying five pathways through which adaptive action can be undertaken by individuals or discrete subgroups within an organisation. These actions are a consequence of interactions with the institutional architecture, adaptive capacity and facility for learning held by an organisation. The host organisation can be any collective social unit, from a legally mandated government department or agency to more flexible but nonetheless restrictive private sector organisations or loose associations of actors from civil society in organised networks.

Feint arrows indicate the direction of conditioning between these aspects of organisational life, such that the institutional architecture conditions the type and scope of learning and adaptive capacity (either directly or through prescribed forms of learning) and adaptive agency through the setting and policing of formal objectives, structures and guidelines for practice in the organisation. Adaptive capacity conditions adaptive actions by setting constraints on what can be thought and done as well as the goodness of fit of existing resources to the identified external pressure. The organisation draws resources, opportunities and threats from the external environment and exerts adaptive actions upon it.

Only two of the five forms of adaptation are visible from outside the organisation. This not only flags the need to examine interior life of organisations but also the overlapping of resilience, transition and transformation with the latter two modes of adaptation potentially being found operating inside an organisation which itself applies only resilience to the external world; indeed transition or transformation might be necessary internal motors for external resilience (see Figure 3.1).

The five adaptive pathways are characterised in Table 3.1. Adaptive agency refers to that held by individuals, sub-divisions or cross-cutting communities of practice below the level of the organisation. Such agency is made reflexive through learning (pathway 1) which is itself an adaptive act. Reflexivity implies strategic decision-making (of an entrepreneurial individual, advocacy coalition and so on within the organisation). This can be focused on symptoms leading to adaptive ingenuity, or consider root causes indicating to critical reasoning. Learning through critical reasoning or adaptive ingenuity can lead to lobbying for change in the institutional architecture of the organisation (2). This can in turn force reconsideration of aims and behaviour informing the selection and use of resources that form adaptive capacity (3) (a key concern given a dynamic external environment). Feedback from capacity and institutional architecture then potentially reshapes adaptive agency.

One important message from this approach to adaptation is that only two modes of adaptation are observable from outside the organisation. These are efforts to shape the relationship between the organisation and operating environment made either by the agent (4) or organisation (5). This is a strong argument for studies of adaptation to climate change to extend their analysis from the external/public to more internal and private life of organisations of all kinds. The

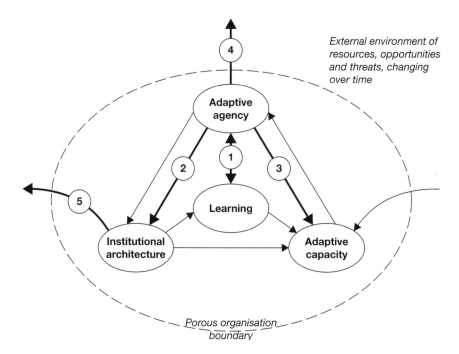

Figure 3.1 Adaptation pathways within an organisation

Table 3.1 Five adaptive pathways

Pathway	Summary	Example
1 Agent-centred reflexive adaptation	Adaptation informed by dutero-learning – reflecting on past actions. This can lead to changes in the practice of learning, management of adaptive capacity, the institutional architecture or directly on the external environment.	A manager decides that existing work guidelines undermine sustainability and so implements reform.
2 Agent-centred institutional modification	The agent undertakes to alter the institutional context within which it operates so as to shift the institutions which control its scope for future adaptive capacity and action.	A scientific advisor lobbies policy-makers to change policy priorities.
3 Agent-centred resource management	The agent unilaterally changes the selection or use of resources to undertake predetermined adaptive action.	While no guidelines exist, a manager adjusts work routines to meet a changing environment.
4 Agent-led external action	The individual agent undertakes adaptive action oriented to the external environment either in compliance with existing institutions or as part of the shadow system; that is, without instruction from the organisation.	Local agents make experimental or spontaneous modifications to physical infrastructure.
5 Organisational external action	The organisation takes action to modify its relationship with the external environment.	The organisation changes its external communication strategy.

model excludes routine responses to external stimuli which are considered part of the legacy of past rounds of adaptation already integrated into existing management and culture.

As has been argued above, self-organised (agent centred), reflexive adaptation targeted at the external environment (4) or institutional architecture (2) are arguably the most significant indicators of resilience. An organisation that enables reflexive adaptation through internal critical reasoning is more likely to be able to respond to abrupt and unforeseen threats and opportunities associated with climate change. Reflexive adaptation, especially that which seeks to challenge existing canonical institutions, is strengthened by a strong shadow system. The key challenge for organisations is how to support – but not to manage – the shadow system. This is a question we turn to in Chapter 6.

Conclusion

Adaptation as resilience is characterised by actions that seek to protect priority functions in the face of external threat. In this way resilience does not directly seek to realign development relations. However, within individual organisations protecting priority functions can require internal transitional or potentially transformational change. Existing literature on resilience has attached a range of meanings to this term. Most developed is the socio-ecological systems literature that characterises resilient systems as those that exhibit capacity for social learning and self-organisation as well as displaying functional persistence. Social learning describes the pathways and social relationships that shape information exchange and can lead to new ways of thinking or acting; self-organisation is attributed to novel and un-directed collective action. Both are generic social phenomenon that can be used to examine the shaping of transitional and transformative adaptation. The distinction between these levels of adaptation is the focus and intention behind social learning and self-organisation rather than the mode.

Organisational theory is complementary to SES in providing a framework to examine the ways in which relationships between canonical and shadow systems within and between organisations shape information exchange, providing scope for learning and innovation. Shadow systems provide a key resource for experimentation because ideas and actions are hidden from formal scrutiny. They can also be a challenge for formal management and so represent a source of conflict between the imperatives for flexibility and transparency. This is an especially difficult problem to resolve in public sector organisations and corporate civil society organisations. Wenger's concepts of boundary organisations and individuals that can work to transfer information between epistemic communities helps to identify a key resource in adapting to climate change where conversations between science and policy are required to prevent maladaptation. They can also help to overcome the limited conceptualisation of climate change adaptation which continues to be framed as a primarily technical rather than a social and political agenda. This framing is in part an outcome of professional specialism and the division of development policy into ministerial silos, and also of the short-term decision-making enforced through budget and electoral cycles. Working to support boundary organisations and to better understand the shadow systems are two ways in which resilience as adaptation can be supported and potentially allowed to contribute towards wider movements of transitional and transformative social change.

4 Adaptation as transition

Risk and governance

When special efforts are made by a diffusion agency, it is possible to narrow, or at least prevent the widening of, socioeconomic gaps in a social system. In other words, widening gaps are not inevitable.

(Rogers, 1995: 442)

Because socio-ecological relations are embedded within economic, political, social and cultural relations, adaptation will touch every aspect of social life, not simply an actor's vulnerability to the impacts of climate change. This will include relations with distant others, to future generations as well as those living in geographically far-away places, now connected by the time–space compressions (Massey, 1994) and teleconnections (Adger *et al.*, 2009b) of globalisation and global environmental change. Rogers' (1995) opening observation reminds us of the intimate connections between the spreading of new ideas and practices in society and social context. Even planned innovation and adaptation in society can exaggerate existing inequalities or generate new ones. Without care, those with most assets and freedom to adapt to climate change will gain additional advantage over those who do not. Rogers' work on the interaction of social and cultural context with the diffusion of innovations in society has wider relevance to studies of adaptation (Atwell *et al.*, 2008). These are returned to in Chapter 9.

The aim of this chapter is to provide a broad conceptual framework to examine adaptation as transition. This is incremental change to social (including economic, political and cultural) relations as part of adapting to climate change. Transitional acts can describe both those that do not intend, or do not result in, regime change (see Chapter 5), but do seek to implement innovations and exercise existing rights within the prevailing order. Transitional adaptation is therefore an intermediary form of adaptation. It can indicate an extension of resilient adaptation to include a greater focus on governance, or an incomplete form of transformational adaptation that falls short of political regime change. From an empirical perspective intent is as important as outcome in indicating transitional, resilient or transformational adaptation. Not all transitional actions will achieve the intended outcomes but they nonetheless reveal critical capacity through intention.

The association of innovation making with rights makes the social dimension of adaptation explicit, even when there is no observed change to regime form.

This is especially true in those social contexts where rights may have been dormant or suppressed under the prevailing social system. Where this is the case, their invocation can generate transitional social change. There are many examples of this from the environmental justice movement where the asserting of existing legal rights is seen as a method for reducing vulnerability to industrial pollution, and at the same time reinforces these same rights (as well as building other capacities in local social organisation, confidence and so on) (Melosi, 2000; Agyeman *et al.*, 2003).

The multiple scales at which social systems operate means that transition can theoretically be implemented and observed operating at the levels of local, community or regime systems, and in more complicated analysis across these scales. Where adaptations and associated rights claims involve multiple levels of actor (for example, a squatter community and municipal government) transitional adaptation demonstrates the importance of multi- or cross-scale analysis (see below).

In this chapter we are interested in the potential for adaptation associated with climate change to open space for wider and connected reform within the constraints of existing governance regimes. This provides a driver for actors to assert rights or claim entitlements to participate in development and risk management and enable progressive transition; that is, improved action on social justice and environmental integrity as part of everyday development.

This chapter extends the framework of social learning and self-organisation put forward in Chapter 3 by considering two additions. First an actor-oriented approach to regime theory is offered. This is then developed by literature on socio-technological transitions to examine in more detail the pressures shaping innovation and the dissemination of alternative visions and practices through governance regimes. The combined framework is then illustrated through an analysis of the opportunities for transitional adaptation in a single management regime rising to prominence in climate change adaptation: urban disaster risk management. This also helps to prefigure the detailed analysis of urban, primarily transitional adaptation presented in Chapter 7.

A vision of adaptation as transition

Transitional action is targeted at reform in the application of governance. It goes beyond the aim of functional persistence but falls short of aiming directly to realign political structures. Both transition and transformation indicate adaptation in governance systems. For transitional adaptation reform is incremental, undertaken at the level of individual policy sectors or specific geographical areas. There is the potential for bottom-up, aggregate transformational change through, for example, the promotion of stakeholder participation in decision-making, leading to the inclusion of new perspectives and values in emerging policy. By contrast adaptation as transformation is composed of adaptive acts that consciously target reform in or replacement of the dominant political-cultural regime as primary or secondary goals (see Chapters 5 and 8).

Governance and transition

Governance systems are composed of multiple actors including public, private or civil society organisations held together through formal and informal institutions that reproduce the balance of power and direction of development pathways in society. Governance operates at multiple scales with overlapping influence; for example, a local water management regime including user groups might be responsible for day-to-day decision-making but given legitimacy by, and made accountable to, national regulators. The nested and overlapping quality of governance regimes provides opportunity for adaptations to spread horizontally and vertically through communication networks and also for the top-down support or dissemination of adaptive capacity and practices. Where progressive adaptation as transition challenges established practices, or those of other competing emergent adaptations, overlapping administrative responsibilities with internal interactions, alliances and linkages can generate resistance to change (Foxton, 2007). Some degree of institutional inertia is healthy for governance systems to provide stability and so a predictable policy environment for development actors to invest. The focus of adaptation as transition is on the shifting balance between stability and reform in the organisational structure played out through movement in the social distribution of exercised rights and responsibilities motivated by perceived and felt pressures associated with climate change. This is articulated through the degree of information dissemination, inclusiveness and influence in decision-making within governance systems and constituent organisations and sectors of practice.

The role of governance in determining the speed and direction of innovation dissemination is illustrated by Atwell *et al.*'s (2008) study of adaptation within the north-central US Corn Belt. They observed that individual farmers' decisions to accept or reject environmentally friendly farming practices were influenced by three scales of governance: individual, community and overarching regime. Interaction between these three layers is further elaborated in transitions theory (see below). The dissemination of new information was most likely to result in the uptake of new practices when communicated to a farmer by a neighbour. Trust embedded in personal relationships and the shared challenges neighbours face was more persuasive than government initiatives and also acted as a sanction on individual actions that might not have been locally socially acceptable as respondents chose between competing strategies in the exercising of farming rights.

All governance systems require actors with varying degrees of power, transparency and legitimacy to undertake at least some limited form of negotiation, causing Young (1999) to describe governance regimes as bargaining processes. The outcomes of negotiation are a product of the relative power of actors during bargaining and the implementation of agreed rules. Change in a governance system 'involves the alteration of the rules and decision-making, not of norms nor principles' (Krasner, 1983: 5). This is an important distinction to make and points research to the possibility that difference and changes within norms or principles over time and space may not be mirrored in changes in the rules and decisions made in the regime. This allows for adaptive changes in administrative structure, technical innovation, land use and so on at the level of governance and its subsequent policy framework without challenging the overarching regime of norms and principles within which governance rests. Over time – or as a result of sudden changes in the operating environment or in internal relations – discontinuities between norms and principles on the one hand and governance mechanisms and practices on the other can potentially trigger transformative change in the regime or top-down pressures for transitional change in the governance system. The extent to which dominant norms and principles and governance rules are antagonistic or complementary may be a good indicator of the resistance of both the governance system and the wider regime to reform including that of progressive adaptation.

A sense of the scope for aspects of governance to enable progressive adaptation to climate change can be indicated by the relationships between actors and institutions. Actors include individuals and organisations with stakes in a policy domain; institutions are for formal (legislation and guidelines) and informal (cultural norms) rules that determine how actors interrelate (North, 1990). Institutions constrain the aspirations and behaviour of actors but can also facilitate change by legitimating processes of critique and reform (Seo and Creed, 2002). Much of the literature on governance change privileges institutions over agency (for example, Krasner, 1983); it focuses on the power of decision-making procedures, rules and cultures to determine scope for reform. The focus is on understanding the persistence of institutions over time rather than how they may be changed and the role of actors in this. For example, Gunderson and Holling (2002) refer to rigidity traps where people and institutions try to resist change and persist with their current management and governance system despite a clear recognition that change is essential.

Emerging actor-oriented approaches offer scope for exploring how institutions come to be changed – a central question in studies of adaptation. Work here focuses on networks of social relations and information flow within governance regimes and their capacity to surface new ideas that may generate individual policy entrepreneurs and the evolution of epistemic communities, where actors from across a governance regime come to share a common viewpoint and can collectively promote change through purposeful advocacy or as a result of collective changes in practice where governance systems are more resistant to change (Hasenclever *et al.*, 1997; Warner, 2003).

The points at which change in the institutional architecture might be expected arises from internal contradictions inherent in the institutional systems of governance. Seo and Creed (2002) propose four contradictions which help identify pinch points, where challenges to existing institutional arrangements might be expected to arise. First, where rules and norms are encompassing and general they may confer reputation rewards but constrain other organisational aims such as efficiency or rent-seeking. This tension encourages selective-decoupling where ritual conformity hides deviations. As the gap between the demands of legitimacy and other behaviour expands so pressure for institutional change grows. Second, path dependency (Arthur, 1989) suggests that incremental investments in physical, human and social capital lead organisations to prefer investments to protect established functions and practices even in the face of environmental change, until a crisis point forces institutional change. Third, where the wide range of institutions within society can lead to conflicts or inconsistencies generated by the interactions between the institutional arrangements of different levels or sectors within a regime forcing change. Fourth, assuming institutions reflect and protect the interests of the more dominant political actors in society, institutional change can arise from political re-alignment and also when previously passive or marginal actors become conscious of the institutional conditions that leave their needs unmet (Benson, 1977). The capacity for critical consciousness and actor reflection on established institutions is arguably the most fundamental element of any actor-oriented governance reform. The extent to which critical consciousness is able to generate actor mobilisation and collective action is explored in more detail in Chapters 6, 7 and 8.

The actor-oriented approach is also useful in avoiding simple, causal explanations for social organisation outcomes. This includes assumptions about the hegemony of state power, the subordination of the local communities and the superiority of the laws of the market (Booth, 1994). In this way one of the significant features of the actor-oriented approach is that it places explanatory value on the agency of even apparently weak or marginal actors (Zimmerer and Basset, 2003). This has encouraged work that has placed emphasis upon the importance of marginal actors (Farrington and Bebbington, 1993) – such as small local NGOs or people at risk. It also allows a more detailed treatment of the state and how it interacts with the non-state groups than was possible under more structural interpretations that saw the state bluntly as a homogeneous entity and also as a tool of the most powerful classes to protect their best interests (Watts, 2000).

The context specific nature of the interaction between actors and institutions is well illustrated by Warner's (2003) comparative analysis of political entrepreneurship in flood management for Bangladesh and the Netherlands. In Bangladesh rigid governance and a conservative administrative culture constrained opportunities for change from within so that governance responded best to pressure exerted by powerful, external actors; in this case development aid donors. In the Netherlands a flexible governance form fostered political entrepreneurship and allowed interdepartmental alliances to form and collectively push for reform from within, although the participatory Dutch administrative culture actually slowed this process through drawn-out rounds of consultation which also acted as an opportunity for thorough review of proposed reforms.

Where might alternative visions and practices be fostered within existing governance regimes, and how might such innovations be tested and diffused to the wider society? Where canonical systems resist change Chapter 3 shows the scope for shadow systems to act as a place of experimentation and learning within organisations. Can shadow spaces exist at the level of the governance systems in collaborations between social organisations on projects that are not supported by or run counter to the dominant governance regime? If yes, can they offer a place for building diversity in thinking and practice that can be formalised if new problems arise (Cohen *et al.*, 1972)? The evidence presented in Chapters 7 and 8 suggests strongly that shadow systems operating as informal networks are a powerful influence of capacity for transitional adaptation. Such polycentric forms can both spread and reduce risk in society, and compensate for failures in other levels of governance (Ostrom, 2005). An example of this is the provision of services or information that are not available formally – information on how to undertake local adaptive measures or by providing post-disaster lifelines when state agencies are compromised (see also IFRC, 2010). Work on socio-technical systems offers some insight onto these questions and it is to this literature that we now turn to add some detail to the broad framework of actor-oriented analysis of regimes.

Socio-technical transitions

Socio-technical transitions work has sought to examine what it is that directs individual development pathways. Reminiscent of coevolutionary theory, transitions in policy or economic domains are explained through changes in and interaction between technological innovation, cultural preferences, industrial production processes, government incentives and demography (Seyfang and Smith, 2007). Insight from this literature is helpful in refining a framework to help understand processes of transition in adaptation, although the emphasis of enquiry moves from delineating histories of socio-technical change to identifying those characteristics of societies that can influence the emergence of opportunities for transition as part of adaptation, the consequences of which might be progressive or regressive for social justice and environmental integrity.

The socio-technical transitional literature, which draws broadly from systems science, is compatible with our existing framework in so far as it acknowledges the role of power (Rip and Kemp, 1998) and agency (Seyfang and Smith, 2007) of competing interests, embodied in innovations, established practices and institutions interacting often across governance scales in shaping the institutional architecture of development. One confounding limitation of this literature is a failure to distinguish adequately between transitional and transformational change. Both are used, sometimes synonymously. At root this is a failure to separate governance systems from the overarching socio-political regime. The former is taken as a sub-set of the latter with transitional change an aspect of transformation and not identified as a goal in itself. Thus, drawing on Rotmans *et al.* (2001), Jerneck and Olsson (2008: 176) are able to claim that 'Transitions and transformation processes in societies, or subsystems thereof, change profoundly in terms of structures, institutions and relations between actors. After a transition, the society, or a subsystem, operates according to new assumptions and rules'.

While on the ground it may not always be clear or helpful to distinguish between transition and transformation, it is nonetheless important to identify transformation as an extreme case where profound change alters the distribution of rights and responsibilities and visions of development across society (see Chapter 5). Individual transitions fall short of this and describe incremental changes to the aims and practices of geographically or sectorally bound activities that push but do not overturn established political regimes. For this reason the transitions perspective is used below to help identify pathways for change. These can then be applied to analyse capacity for, or past trajectories with, transitional (claiming rights within existing regimes) or transformational (replacing regimes with new rights compacts) outcomes. Empirical work presented in Chapter 7 applies this framework to examine transitional adaptation and its messy connections with resilience and transformation in the Mexican Caribbean.

From its origins in industrial history, the transitions literature has expanded to critiques of sustainable development, arguing that not only contemporary practices but the solutions they generate for development challenges are unsustainable because they fail to address fundamental values driving dominant development paths (Rotmans *et al.*, 2001). Transitions framings have most recently been applied to climate change mitigation – through, for example, research on transition to low-carbon living in the UK (Haxeltine and Seyfang, 2009). Both agendas demonstrate the potential utility of transitions framing for contributing to understandings of transitional (and transformational) adaptations. This is particularly so in the context of poorer societies at risk to the local impacts of climate change where adaptation is a more pressing priority than mitigation and extremes in social inequality and access to human rights and basic needs demand adaptation confront failed development policy regimes.

Like the socio-technical transitions proposed for sustainable development challenges, analysing opportunities and outcomes for a progressive, transitional adaptation benefits from a lens that can examine technological innovation and evolution as a social process. Geels (2005) describes the protected spaces where new technological or management innovations develop as socio-technical niches, suggesting that innovation originates from local experimentation. Niches are set in contrast with the dominant socio-technical regime operating at the meso-level, and the larger macro-level contextualising political, economic, cultural and environmental 'landscape'. Berkhaut *et al.* (2004) also observe that change in the regime may be driven top-down by perturbations in the wider landscape. Amongst the most useful findings of this literature are proposals, supported by case study work, for specific pathways and strategies that determine how far a socio-technical innovation is able to escape from its protective niche and overturn the dominant regime (Geels and Schot, 2007), and the identification of the barriers to regime change that promote path dependence. These barriers include the repetition of cognitive routines that blind professionals to developments outside their focus (Nelson and Winter, 1982), regulations and standards that enforce rigidity (Unruh, 2000), lock-in of adaptation to technical systems so that the costs of transition relative to resilience increase over time through fixed investments in machines, infrastructures and competencies – until systems thresholds are crossed by external drivers such as regulation, market changes or natural disaster (Tushman and Anderson, 1986). These drivers and constraints have suggested to some analysts that activity at the niche level alone is not enough to generate transitions and rather this is an outcome of multi-level collaboration and in the process local experiments and regime practices will be mutually adjusted and compromised (Seyfang and Smith, 2007).

The closest empirical focus to adaptation has perhaps been made by Seyfang and Smith (2007) who recalibrated socio-technological transitions theory to examine the emergence and diffusion of ideas and practices from grassroots environmentalism in Europe including social businesses, cooperatives and informal community groups. This is helpful in drawing out differences between entrepreneurial innovation, where risks and rewards are managed through the market with modest government support; and innovation in the social economy, where local groups and individuals invest social as well as economic capital in innovations and where rewards tend to be more dissipated and motivations are driven more by ideology than economics. Grassroots niches are found to be catalysts for participation where individuals and communities can build confidence, a shared sense of values and ambition for change, skills and capacity for further rounds of learning and innovation. Some of these very assets that make communities adapt at innovation can also be barriers for diffusion including the geographical specificity of experiments, strong identity and visions for the future which may be antagonistic with overarching regimes suggesting that some degree of translation and compromise is required for bottom-up reform, or that there is significant change affecting the regime from the top down to motivate the search for alternatives (Church, 2005). This is precisely the opportunity that climate change brings.

Haxeltine and Seyfang (2009) identify three modes through which local innovations can come to influence the wider regime:

- *Replication*: horizontal reproduction through multiple, small initiatives
- *Scaling-up*: the expansion of individual initiatives as they attract more participants
- *Mainstreaming*: the absorption of innovations into dominant policy and practice.

Experience from attempts at spreading disaster risk reduction from local community initiatives shows just how difficult grassroots socio-technical transitions can be. This is the case especially when the dominant regime is ambivalent, sceptical or antagonistic, and where innovation is perceived to threaten local or national established interests. Challenges arise when both seeking to support local innovation without disrupting preferred local social relations, and in promoting the wider influence of successful innovations. Box 4.1 summarises a review of experience generated through a workshop discussion with community-based disaster risk managers including international agencies such Tearfund,

Box 4.1 Lessons in making transitions from community-based disaster risk management

Community-based disaster risk management fosters inclusive approaches to disaster risk reduction, with a special focus on livelihood sustainability in light of disaster risk, and calls for the utilisation of local knowledge and skills. It has become a popular strategy for international and local development and humanitarian NGOs seeking to find ways of empowering those at risk to manage their own vulnerability. Despite such enthusiasm there are relatively few examples of long-term success. Participants in the workshop identified nine reasons for this:

- Ineffective translation and communication of climate science within communities.
- A lack of long-term financial support for local capacity-building and the longitudinal application of community-based disaster risk management.
- Local abuse of power granted by community-based disaster risk management to local actors, which can lead to community fragmentation.
- Resistance from local elites when community-based disaster risk management is perceived as a threat to the status quo.
- Difficulties for implementation in unstable communities facing economic or political stress.
- Difficulties for implementation on a larger scale, and related limits on the analytical and policy applications generated by community-based disaster risk management.
- A failure to link with broader and/or longer-term development priorities and activities.
- The generation of inequality as some neighbourhoods build resilience through community-based disaster risk management while others remain vulnerable.
- The lack of local success stories to act as examples for scaling-up and replication.

These challenges to the repeating of success are likely to apply across policy fields, to other contexts where progressive external actors seek to build local adaptive capacity through empowerment methods. Specific challenges were also reported from those actors who had attempted to support or lead mainstreaming, scaling-up and replication of local successes.

Mainstreaming in governments and communities faced distinct challenges. For governments, seeking support for community-based disaster risk management and its outputs led to competition with other priorities and required a framework to link efforts between local and national actors.

Local communities needed institutional channels to establish dialogue with the government, particularly about risk perception, diversity within the community (social and economic) and legal entitlements to risk reduction.

Scaling-up required a functioning institutional infrastructure and so was most likely to be found within a supportive, inclusive and open governance system.

Replication was also be found where central institutions were supportive but did not rely on this. It can provide a means of reproducing good local practice when governance systems are unable or unwilling to support scaling-up. In urban slum settlements and isolated rural communities beyond the reach of the state, expansion through horizontal replication was undertaken where existing networks of community actors and organisations could provide the institutional framework for replication.

Community-based disaster risk management was found to be most successful when local actors, local leaders and government representatives led and worked together. Where this was possible, it maximised opportunities for mobilising joint action. Furthermore, partnerships led to additional benefits from the extension of local resources and the building of generic human capacity as local actors turned from beneficiaries to planners and advocates taking on rights and responsibility for local risk management.

Where there was an institutional framework to support bottom-up initiatives, this enabled local actors to feed into analysis and interventions designed to address structural issues related to national-level vulnerabilities. These plans were otherwise beyond the reach of community-based disaster risk management and at times conflicted with it. The major challenge for reproducing community-based disaster risk management was how to successfully encourage sustained governmental involvement set against competing budgetary pressures and with the potential that bottom-up innovation may challenge existing norms and practice.

Source: ProVention Consortium (2008)

Oxfam and the Red Cross, and networks of local development organisations such as GROOTS and local NGOs. All had experience of attempting to reproduce the success of pilot projects that had asserted the rights of local marginalised actors (predominantly the poor and women). All were also the product of local action inspired from the outside and top-down (albeit from progressive civil society actors) and were not initiated locally. The lack of local innovation is a key aspect of vulnerability indicating a lack of adaptive capacity. It is also a challenge for external agencies that seek to facilitate local actors in the building of capacity. Experience indicates this is a long process requiring generational shifts in attitudes and individual identity. But extremes of poverty and now also the urgency of climate change put pressure on progressive, external actors to accelerate this process.

An important message from Box 4.1 is the difficulty experienced by external actors seeking to stimulate innovation and diffusion. In low-income countries where the market has limited reach only the state has the scope and resources to spread innovations throughout society, but is resistant to change. In this context, transitions unfolded over time and with uncertain trajectories, the majority of innovations never extend beyond local impact. For those that did, at some point transitions became coordinated either strategically through a lead actor, like the state agency, or as a convergence of the visions and actions of diverse groups (Geels and Schot, 2007) analogous to a social movement.

The interaction of local innovation with the wider regime in shaping transition is a repeated theme summarised by Geels and Schot (2007) into five transition pathways (see Table 4.1). Each is a specific outcome of the interaction between local innovations and the wider regime. The pathways have been renamed in Table 4.1 to better draw out the salient characteristics for transitions in regimes.

Climate change acts as an external pressure on the regime through changes in markets, international regulation, aid and trade flows as well as environmental risk. Local adaptations can then potentially be inserted into the regime to meet these new challenges as a means of strengthening the status quo following moderate impacts (weak cooption), or after catastrophic change in a search for the realignment of the regime to a new external environment (innovative substitution and innovative competition), until a new round of challenges emerge. The framework was designed to account for change in broad patterns of production and consumption. In so doing it arguably over-emphasises the role of top-down pressure and external triggers and underplays the internal competition between

Table 4.1 Transition pathways

Pathway	Characteristics
Stability	In the absence of external and internal pressures there is limited scope for local innovations to affect change in the regime, though they can act as a resource against an uncertain future.
Top-down reform	Moderate external pressure is acted upon by regime actors rather than local innovators to change the direction of regime policy and practice from within.
Weak cooption	Unchallenging local innovations are incorporated by the regime but their adoption triggers unforeseen adjustments to the regime.
Innovative substitution	Catastrophic external pressure shows the regime to have failed. A single innovation has already been developed and can be inserted into the regime.
Innovative competition	Catastrophic external pressure shows the regime to have failed. Multiple innovations compete for dominance until a new stability is achieved.

multiple viewpoints on development that persist even during periods of perceived stability and can rise to challenge existing institutions and regimes.

There are parallels between transitions theory and the resilience framing of adaptation as discussed in Chapter 3. Both identify critical moments for adaptation as a tension between innovation in sheltered spaces, be it niches or the shadow systems; and subsequent efforts to influence across the system of interest, be it the organisation or governance regime. This is not surprising: both approaches draw on system theory differing in the scales and contexts of application. In organisations and in wider society potential for capacity innovation is indicated through learning processes, social networks, communities of practice and advocacy coalitions that seek to modify institutional structures to allow wider diffusions of innovations (Pelling *et al.*, 2007; Smith, 2007). Both point to a creative optimum where top-down resources such as political will, financial and technical support are available but without undue oversight so that local actors can be left to experiment, even to take risks in doing so, with the benefit to the wider system of generating an array of ideas and practices. The necessity for a coalition of local and higher-level organisations and interests to enable diffusion is a central message from these literatures, demonstrating the limits of autonomous and spontaneous adaptations.

Urban regimes and transitional adaptation

This section aims to illustrate how spaces for transition might open in adaptation processes within a single governance regime: urban risk management. There are as many different models of urban risk management as there are examples (see, for example, UN-HABITAT, 2007), but some general observations can be made that are also illustrative of the interaction of actors, structures and visions of development to be found in other types of regime and this is the aim here.

The starting point for assessing scope and barriers for transition in adaptation is to recognise the contested social construction of the vision, associated priorities and subsequent practices that give substance to the regime, and also describe those fault lines that are likely to come under pressure and may be realigned as transition unfolds. Competing visions of the city are underlain by ideological, material and economic interests (Kohler and Chaves, 2003). The balance of influence accorded to individual visions determines what urbanisation means, who the winners and losers are – and under adaptation to climate change what aspects of urbanisation are to be protected or are dispensable for any one settlement. Indeed in different places across the city different visions and associated actions will have more or less traction even within a single policy sector. This provides opportunities for alternative experiments that may come to dominance once governance space is opened following a disaster event, political or economic change or macro-administrative decision such as decentralisation.

A range of visions that can provide narratives for the direction of dominant risk management decisions in a city are shown in Table 4.2. Different visions of urbanisation include the city as a motor for generating macro-economic wealth,

Table 4.2 Linking visions of the city to pathways for managing vulnerability

Vision of the city	Vulnerable objects	Pathways for managing vulnerability	Literature
An engine for economic growth	Physical assets and economic infrastructure	Insurance, business continuity planning	Econometrics of business continuity and insurance
An organism or integrated system linking consumption and production	Critical/life-support infrastructure	Mega-projects connecting urban and rural environmental systems	Political-ecology, systems theory
A source of livelihoods	The urban poor, households, livelihood tools	Extending and meeting entitlements to basic needs	Livelihoods analysis and medical sociology
A stock of accumulated assets	Housing and critical/life-support infrastructure	Safe construction and land-use planning	Political-economy and urban sociology
A political and cultural arena	Political freedoms, cultural and intellectual vitality	Inclusive politics and the protection of human rights	Discourse analysis and public administration/ political theory

(Source: Pelling and Wisner, 2009)

an organism turning raw materials into products and waste, a source of livelihoods for urban citizens, an historical accumulation of physical assets and infrastructure or a place for cultural and political exchange and debate.

Visions and associated risk management preferences need not be mutually exclusive and more often visions provide only a rationale for prioritisation. As climate change and other development dynamics interact in the city existing narratives will be tested. With this comes the possibility of opening space for renegotiating priorities within a single policy area and also more broadly so that resources and political will may be drawn into or away from risk management. The complexity of risk management means that there are many potential actors with a stake in the existing regime, and with an interest in any renegotiation of the balance of priorities in risk management that adapting to climate change may offer – whether through replication, up-scaling or mainstreaming. Table 4.3 summarises the kinds of actors likely to have a direct stake in adapting urban risk management sector. The key message to take from this is the wide range of urban actors that can be engaged with through even a single sector, extending from applied emergency and risk management to development regulation and planning. This indicates both the number of opportunities that can exist for

Table 4.3 Urban disaster risk reduction: multiple activities and stakeholders

Professional community	*Development planning*	*Development regulation*	*Risk management*	*Emergency management*
Core activities	Land-use, transport, critical infrastructure	Building codes, pollution control, traffic policing	Vulnerability and risk assessment, building local resilience	Early warning, emergency response and reconstruction planning
Primary stakeholders	Urban planners, city engineers, critical infrastructure planners, homeowners, private property managers, investors, transportation users, taxi drivers' associations, other professional associations, academia	Environmental regulation, law enforcement, contractors, factory owners, drivers' and transporters' associations	Primary health care, sanitation and water supply, community development, local economic development, infrastructure management, waste haulers' association, water users' representatives	Environmental monitoring, emergency services, civil defence, disaster management coordination, fire fighters, police, military, Red Cross/ Crescent society

(Source: Pelling and Wisner, 2009)

inserting progressive practices such as inclusive decision-making or downward-accountability into policy reform in this single sector but also the challenge of overcoming institutional lock-in and inertia that has often been found in sectors with multiple actors where existing inter-organisational alliances act to make the existing regime resilient.

Opportunities exist for innovation niches within private, public and civil sectors and through communities of practice that cross these boundaries. Through local branches or by advocacy at the landscape level international agencies and other governments may also be active in shaping niche or landscape led innovation. Where innovation is fostered and how it is evaluated and diffused throughout the regime in transition will also be a function of the balance of formal and informal or shadow systems in the city. Social networks and communities of practice that cross-cut formal organisational boundaries will influence scope for the development of novel adaptations, and the speed and direction of diffusion and reform. In urban contexts local government should play a pivotal role as a mediator and facilitator of development in addition to any direct regulatory and management roles. That the reality of local government is so often as an under-resourced actor, lacking in human capital, popular legitimacy and political influence is a great constraint. This is illustrative of Jerneck and Olsson's (2008) observation that the regime level tends to be resistant to change and that innovation comes more often from local niches or the wider landscape and is accepted only when the regime is stressed.

Conclusion

Opportunity for transition opens when adaptations, or efforts to build adaptive capacity, intervene in relationships between individual political actors and the institutional architecture that structures governance regimes. Transitional adaptation falls short of directly challenging dominant cultural and political regimes, but can set in place pathways for incremental, transformational change. Both actor-oriented regime theory and socio-technological transition theory provide ways forward for drawing out the connections between adaptation and social evolution short of regime change. They share an emphasis on the role of agency in the dialectical relationship between actors and institutions that constitute governance systems; agency that if fostered by a supportive governance regime can be a resource of alternative ideas and practices available for implementation following changes in the wider physical and economic–political environment.

Existing regimes can block progressive adaptation at the level of transition, especially when change threatens established power relations. Marginalised actors can also be reluctant to undertake change when there is uncertainty over the outcomes of reform or where short-term transactions costs are high, compared to adapting through resilience. Where local adaptations are successful and open transitions, diffusion into the wider society is also challenging without government support. Fragmented transitions run the risk of exacerbating inequalities as successful adaptations not evenly applied across communities or sectors. External shocks that show the existing institutional architecture wanting can potentially provide the impetus needed to generate political will for transition, and potentially also transformation. This is the opportunity that climate change brings.

5 Adaptation as transformation

Risk society, human security and the social contract

Instead of destroying natural inequality, the fundamental compact substitutes, for such physical inequality as nature may have set up between men, an equality that is moral and legitimate, and that men, who may be unequal in strength or intelligence, become every one equal by convention and legal right.

(Rousseau, 1973, original 1762: 181)

For Rousseau, a just society is one where those with power are held to account over their ability to protect core and agreed-upon rights for citizens. As a normative theoretician, Rousseau argued that the ideal social contract, one that confers upon rulers the legitimacy to retain and exercise power, would ultimately be granted by the citizenry, not assumed or god-given: an agreement ratified at the level of culture as well as law, and one that can be transformed if either side fails in its part of the contract. But Rousseau also recognised that the social contract could be undemocratic, imposed with force or through the manipulated complicity of citizens themselves. When prevailing social relations are a root cause of vulnerability and a target for adaptation, this observation means that change will not be easy (Williams, 2007). The classical formulation of the social contract, such as that offered by Rousseau, is also revealing for what it does not include. Rights are extended only to citizens. The globalised and teleconnected impacts of climate change and adaptation decisions require that future generations and those living beyond national boundaries also be considered, as well as the non-human.

This chapter builds on the preceding discussions around adaptation as resilience and transition. These introduced the notions of social learning, self-organisation, actors in regimes and pathways for socio-technological transition. The notions of risk society, the social contract and human security are offered as theoretical devises to help reveal the fault-lines of dominant society. These are by no means the only theoretical lenses that could be brought to help examine transformational adaptation. They have been selected because together they provide a continuum for transformational adaptation that stretches from conceptualisations of development under modernity to the application of policy for national and human security. In this way they provide a landscape of ideas to help position and understand adaptations that seek to address root causes and leverage transformation. Like resilience and transition, transformation can be seen as an intention or as an outcome of adaptation. It also operates at all scales, from the local to international, often simultaneously and in ways that are difficult to perceive. In identifying the assumptions that underlie modernity as a potential focus for adaptation transformation is also directed towards internal – cognitive change; for example, through the production and reproduction of dominant cultural perspectives that emphasise and justify individualism and undermine social solidarity and collective action: a frequently identified key component of local adaptive capacity (Smith *et al.*, 2003).

A vision of adaptation as transformation

The notion of a social contract is not only abstract, it can help in the analysis of crises of legitimacy that precede political regime change, and potentially be used to avoid such crises. Disasters associated with climate change triggers are but one driver of crisis, and do not guarantee transformational change (see Table 5.2). In such cases loss of legitimacy is to be expected when observed risk or losses exceed those that are socially acceptable. Beneath this the consequences of climate change are accepted as a play-off against other gains. Of course not everyone in society will hold the same tolerance to risk or loss and both will change over time as cultural contexts evolve. In this way the social contract is kept in a tension by risk and loss (as well as opportunity) associated with climate change, and also by whose values are included in the social contract. In addition to the established social divisions along lines of class, gender, cultural identity, productive sector, geographical association and so on, climate change also requires the recognition of future generations and distant interests in local decision-framing (O'Brien *et al.*, 2009). The inclusion or exclusion of these voices will determine the extent to which climate change is perceived to con-tribute to individual disasters or crises, and the points at which different actors are held responsible for the management of climate change and its consequences. This in turn shapes priorities for social responses to climate change risk and loss. That the interests of future generations or citizens of second countries should be allowed in this conversation fundamentally challenges established social organisation based upon the nation-state.

Can adapting to climate change incorporate this dynamic and be a mechanism for progressive and transformational change that shifts the balance of political or cultural power in society? Evidence for the potential of transformational change within national boundaries can be found in the slow and limited accept-ance of international aid by the government of Myanmar following Hurricane Nargis. In large part this behaviour was a result of fear of the destabilising influences of international humanitarian and development actors on the regime, a policy that analysts have also attributed to the need for the ruling military elite to demonstrate its control over society – especially at a time when the impacts of the hurricane meant its popular and international legitimacy was at crisis point; and the potential for usurping rich agricultural land from Karen ethnic minority farmers in the Irrawaddy delta where the hurricane made landfall (Klein, 2008). Distrust by the Myanmar regime of international and especially Western and civil society actors has had a byproduct of catalysing organisational reform at the regional level. The leadership of the Association of Southeast Asian Nations (ASEAN), a regional economic grouping, in responding to Hurricane Nargis has resulted in tighter regional cooperation for disaster response. This is an important regional adaptation, one based on a principal of political non-intervention and so a form of adaptation that adds resilience to the status quo. The durability of this position amidst calls for a more engaged approach of 'non-indifference' (Amador III, 2009) and its consequences for bottom-up transitional or transformational capacity are yet to be seen.

Where transitional adaptation is concerned with those actions that seek to exercise or claim rights existing within a regime, but that may not be routinely honoured (for example, the active participation of local actors in decision-making), transformational adaptation describes those actions that can result in the over-turning of established rights systems and the imposition of new regimes. As with adaptation as resilience or transition, any evaluation of the outcome of transfor-mational adaptation will be dependent on the viewpoint of the observer (Poovey, 1998). Efforts undertaken to contain or prevent scope for transformational adaptation are as important as the adaptive pathways themselves in understanding the relationships between climate change associated impacts and social change. For example, it is very common for the social instability that follows disaster events to be contained by state actors. This is achieved through the suppression of emergent social organisation and associated values halting the growth of alternative narratives or practices that might challenge the status quo, and lead to transformation as part of post-event adaptation (Pelling and Dill, 2006).

The socio-ecological systems literature has less to say about transformation than resilience and transitional change. Nelson *et al.* (2007: 397) describe trans-formation as 'a fundamental alteration of the nature of a system once the current ecological, social, or economic conditions become untenable or are undesirable'. But for many people, especially the poor majority population of many countries at the frontline of climate change impacts, everyday life is already undesirable and frequently often chronically untenable. Here we come to a central challenge for systems analysis which places the system itself as the object of analysis. Resistance in a social system can allow it to persist (be resilient) in the face of manifestly untenable or undesirable ecological, social or economic features for sub-system components. Theoretical work on nested systems allows some purchase on this (Adger *et al.*, 2009a), but is very difficult to develop empirically. The points at which these failures lead to challenges for the overarching regime serve as tipping points for transformation. Tipping points that Nelson *et al.* (2007) point out can be driven by failures that are absolute (untenable) or relative (undesirable), so that cultural values play as much a role as thermodynamic, ecological or economic constraints on pushing a system towards transformation.

There is scope for transformation to arise from the incremental change brought about by transitions (see Chapter 4). Subsequent claims on the existing system results in modifications at the subsystem level. Over time and in aggregate this forces an evolutionary transformation in the overarching system under analysis. It is this pathway to transformation which existing climate change literature has focused upon. With an interest in practical ways in which productive systems might transform under climate change, Nelson *et al.* (2007) describe this process as systems adjustment and include the implementation of new management decisions or the redesigning of the built environment as examples. This aspect is considered in Chapter 4.

Classifying transformational adaptation is sharpened by identifying: (1) the unit of assessment – sub-systems and overarching systems may be undergoing different kinds of adaptation, or none at all – as well as their interactions; (2) the viewpoint of the observer, which can place a logic for a normative assessment of transformation; and (3) distinguishing between intention, action and outcome. A single type of action, for example greater local actor participation in risk management decision-making, can promote resilience, transition or transformation – it is the outcome that counts, and outcomes cannot always be planned for. It is the fear of surprises and incremental change in social relations that encourages tight control of emergent social organisation for risk and impact management and forces many actions into the shadow system of informal relations and organisation.

Where should one look to reveal the challenges and potential directions for transformative elements of adaptation? Most practical work on adaptation focuses on addressing proximate causes (infrastructure planning, livelihood management and so on). Transformation, however, is concerned with the wider and less easily visible root causes of vulnerability. These lie in social, cultural, economic and political spheres, often overlapping and interacting. They are difficult to grasp, yet felt nonetheless. They may be so omnipresent that they become naturalised, assumed to be part of the way the world is. They include aspects of life that are globalised as well as those that are more locally configured. The former do not have identifiable sites of production and require individual and local as well as higher order scales of action to resolve (Castells, 1997). The latter are more amenable to action within national and local political space. Table 5.1 identifies three analytical frames that each reveal different aspects of domination and the associated production of vulnerability. Each points to specific indicators for transformation as part of adaptation.

The indicators of transformation identified in Table 5.1 require deep shifts in the ways people and organisations behave and organise values and perceive their place in the world. Together they help describe the features a sustainable and progressive social system might be expected to exhibit. They operate at the level of epistemology: the ways people understand the world. Surface – transitional – changes are already observable; for example, in the uptake of socio-ecological systems framing in adaptation and more widely in natural resource management. But transformation speaks to much broader processes of change that encompass individuals across societies, not only specific areas of professional practice, though such enclaves may yet prove to be the niches that lead to

Table 5.1 Adaptation transforming worldviews

Analytical frame/thesis	Root causes of vulnerability	Indicators of transformation
Risk society	Modernity's fragmented worldview; dominant values and institutions are coproduced at all scales from the global to the individual	Holistic, integrated worldviews including strong sustainable development and socio-ecological systems framing of adaptation and development; adaptation that draws together the value systems of individuals with social institutions
Social contract	Loss of accountability or unilateral imposition of authority in economic and political relationships	Local accountability of global capital and national governments, to include the marginalised and future generations and not be bound by nationalistic demarcations of citizenship
Human security	National interests dominate over human needs and rights	Human-centred approach to safety, built on basic needs and human rights fulfilment, not on militarisation or the prioritising of security for interests in command of national level policy

profound societal change (see Chapter 4). More tangibly, transformation that moves beyond intention also unfolds at the level of political regime. Here the root causes of vulnerability are made most visible when latent vulnerability is realised by disaster. The post-disaster period is an important one for understanding the interplay of dominant and alternative discourses and organisation for development and risk management and is examined below.

Following this initial discussion of the nature of transformational adaptation this chapter examines risk society, human security and the social contract as lenses to direct analysis of transformation within adaptation. The influence of disasters as moments in national political life that can catalyse regime change are then reviewed.

Modernity and risk society

For Western science and policy discourse, fear of surprises from climate change has been predominantly interpreted through adaptation as requiring actions that can help to manage risk by greater control of the environment. This confounds proximate with root causes of climate change risk (Pelling, 2010). Environmental hazard under climate change is an outcome of the coevolution of human and bio-physical systems, not simply of external environmental systems acting upon human interests. In this context, perhaps the most profound act of transformation facing humanity as it comes to live with climate change requires a cultural shift from seeing adaptation as managing the environment 'out there' to learning how to reorganise social and socio-ecological relationships, procedures and underlying values 'in here'. This in turn demands a strong component of conceptual and social reorganisation. How far this might precipitate political and economic regime change is unclear.

Ulrich Beck has written extensively on the nexus of modernity, technology, the environment and human security in what he calls risk society (Beck, 1992). His theory of reflexive modernity posits that transformations in the nature of rationality are the basis for contemporary environmental and social challenges, and it is at this deep level that change must arise for risks to be avoided at root cause. Beck argues that modernity has produced a simplified model to understand the world, one that fragments and isolates different components. This approach has led to the application of sector specific technologies for development and risk management. There are undeniable successes with this approach but climate change impacts reveal the limits. The complexity of socio-ecological systems dynamism exemplified by climate change and adaptation cannot be captured by individual sectors or sciences. The influence of socio-ecological thinking and systems theory in the sciences is one response to this. Policy actors have been slower to respond with administrative structures and political regimes, sticking with increasingly inappropriate structures. These are structures – in administrations including Ministries of the Environment, Civil Defence, Central Banks and Foreign Affairs – that need to work together to adapt progressively to the risks of climate change. Beck's analysis is striking, suggesting that the isolated and fragmented nature of management and practical technologies created within this model of reality allow uncontrolled interactions inbetween. This results in unforeseen and catastrophic consequences including climate change. Moreover, due to the closed nature of the system, alternative trajectories are blocked:

> Risk society arises in the continuity of autonomous modernization processes, which are blind and deaf to their own effects and threats. Cumulatively and latently, the latter produce threats, which call into question and eventually destroy the foundations of industrial society. (Beck, 1992:5–6)

While developed with rich, industrialised economies in mind, Beck's basic thesis is transferable to those poorer countries that have used industrialisation to drive agricultural and urban development (Leonard, 2009). Foreign direct investment and the conditionalities of aid that promote market led industrialisation from the outside broaden responsibility and perhaps undermine the reflexivity of local and even natioanl actors. But the root causes of the climate change challenge and the consequent need to situate adaptation (and mitigation) within development, not as technical adjucts to it, remains the same. Beck channels his hope for recovery in the formulation of a radically alternative model of modernity, but finds the tools and insights needed to challenge the socioeconomic policies that lead us towards disastrous outcomes *within* society. He believes that risk society is inherently reflexive and perceives the contradictions between its original premises (human advancement), and the outcomes (environmental disaster), but argues that radical change must come from socio-political interventions designed to transform development driven by industrialisation, going well beyond the risk management agenda, including that being associated with climate change. Progress has been slow; five years after publishing his thesis of the relationship between risk society and modernity, Beck observed that consumption had become a key driver alongside industrialisation in the production of hazards and vulnerabilities. This suggested that neoliberal economic policies are an increasing threat to human security – reflexivity has not yet led to transformation but rather an acceleration of the root causes of crisis (Beck, 1999). In the succeeding decade little has changed beyond the increasing pace and intensity of consumption and associated risk production, and the depth of inequality in risk at scales from the global to local. A slow but growing concensus for the reorganisation of technology and finance was given a brief and short lived stimulace by the 2008–10 global economic crisis.

In the existing policy landscape, the challenges of attaining a more holistic and reflexive approach to living with climate change and its impacts can be seen playing out in the disaster risk reduction community – a critical component of adaptation. This community has long argued for the advantages of conceiving of environmental risk as embedded within development and of confronting development, not the environment, in seeking to reduce risk, but has a long way to go in embedding this within dominant policy frameworks. Some progress is being made and is reflected, for example, in sections of the guiding Hyogo Framework for Action 2005–15. This international agreement is non-legally binding but compels signatory states to work towards five areas of action. The first of these promotes institution building and calls for the integration of disaster risk management within development frameworks such as poverty reduction strategies (see http://www.unisdr.org). Reflexivity, though, is ultimately about changing values and requires a political and cultural process in addition to the sectoral-technical one described above.

The social contract

The idea of a social contract is the foundation of contemporary liberalism, underwriting both its Kantian and Utilitarian strains (Hudson, 2003), and has had an enormous influence on the construction of liberal political ideologies and institutions (Pateman and Mills, 2007). The social contract is useful for an analysis of transformational adaptation and its limits because it draws attention to the compact in society that determines responsibility for risk management as part of development. More than this, it helps to reveal the basis of legitimacy of this understanding and so the potential fragility of the existing status quo. This can be examined, for example, through exposing the balance of market, public and social pathways through which security is framed and mediated. Issues of responsibility and pathway are open for contestation as failures in the social contract are revealed during disaster and its aftermath (Pelling and Dill, 2006). Recently, for example, concerns have been raised about the increasing role of the international private sector in disaster reconstruction and subsequent loss of accountability to local actors at risk, and also to those citizens, often of second countries, who provide funds through tax payments of charitable donations (Klein, 2007).

The social contract is a product of Western liberal philosophy. However, the generic idea of collective understanding in which parties agree to cooperate with one another, seeding power according to a set of rules, is not. This can be seen in traditional human understandings of rules concerning socio-cultural and biological reproduction, exchange, reciprocity and respect (Osteen, 2002), recognising the parallels to the social contract operating at different scales and in contrasting cultures where the state is not the dominant actor opens useful scope. This can help in extending the focus of that aspect of social contract theory that explores the often tense and sometimes dynamic distribution of rights and responsibilities in society to other interactions than those between the nation state and its citizens.

In the tradition of Western political philosophy the notion of a social contract has not been built on observed customary practice but rather an abstract set of ideas about the nature of political authority and popular consent (Gierke, 1934; Buckle, 1993) stretching back to the work of Thomas Aquinas (circa 1250). Hobbes, Rousseau, and Locke each developed versions of social contract theory. They shared some similarities in approach, drawing from the idea that each human is born into a state of nature and endowed with absolute equality, but with no protections whatsoever against the unregulated violence of anarchy existing where each human competes with all others. They theorised the social conditions under which people might engage as stakeholders in a political society to mediate the violence of an anarchic society. The social contract was the basis for the creation of political societies in which all could secure their basic needs, exercise creativity and enjoy individual autonomy in peaceful sociality. But since the trade-off involved ordinary people forfeiting all or some amount of their freedom/power to the dominant social actor (the state) in order to ensure personal security, the derivation of state authority (previously understood as divine right) was suddenly understood as originating in a consensus of the people. These were deeply radical ideas conceived during a period when absolutist governance and feudalism were destabilising in Europe. Social contract theory constituted the philosophical counterpart to the political and economic changes occurring during the transition to modernity in Europe.

These early theories of state and society paved the way for the creation of the liberal political tradition spearheaded by Adam Smith and Jeremy Bentham. Ironically, it was the theoretical descendents of Hobbes rather than Locke that argued most forcefully for a restricted state. Jordan (1985) explains that like Hobbes, the Utilitarian branch of liberal thought viewed essential human nature as seeking pleasure and avoiding pain, but the Hobbesian strong state (designed to ameliorate the conflicts that these motivations would inevitably provoke) was rejected. Instead, Bentham and James Mill subscribed to the view that since governments were made up of humans who would attempt to enrich themselves and seek to increase their power over their subjects, the power and reach of government had to be restrained.

Debate on the most appropriate balance of power between the state and civil society (including the market) continues to this day (van Rooy, 1998). In a precursor to post-modern thinking, Gramsci (1975) argued that the division between political society and civil society was artificial. Just as hegemony captures the state and civil society, and associated fields of culture and education, so too the counter-narratives of belief systems can be found cross-cutting the divides of the market, state and civil society. Consequently, for Gramsci, while the benefits of the social contract extended only as far as the bourgeois periphery, its universalist language is produced and disseminated across all society so that those subservient to the social contract are also caught up in its reproduction. This accounts for some of the inertia in political regimes where inequality is made manifest through disaster and reconstruction yet where pressures for change in the social contract fail to attain popular support. Gramsci believed that by offering marginalised populations the tools of critical thinking, and the structure of organised groups to bring their distinctive cultures to bear in the production of counter-hegemonic discourse, transformational change could be achieved, at least at the level of discourse (Urbinati, 1998).

Habermas (1976) offers a second possibility for transformation through a crisis of legitimacy. This follows the failure of the dominant actor in the social contract to meet its own responsibilities. In this understanding new critical awareness is not required to make the failures of the social contract visible. But Habermas does argue (1985) that collective action is a necessary condition for realising social and political change once the failings of legitimacy are revealed. Both Gramsci and Habermas place great emphasis on the role of culture and identity, and the influence of education on this, in demarcating the fault-lines along which the social contract is vulnerable to transformation, and also its resilience. Identity that is ascribed to social markers such as race or ethnicity (Hite, 1996), but whose logic also extends to include identity through association with place (Wagstaff, 2007) is significant for understanding the transformational possibility of environmental crisis which has the power of physical destruction. It is the potential for disaster to destroy social life (Hewitt, 1997) and the cultural meanings invested in the physical, as well as physical assets themselves, that in turn opens scope for new understandings of identity and social organisation that offer an alternative to established structures in the social contract when legitimacy is lost (Pelling and Dill, 2010).

The application of social contract theory to questions of climate change resilience draws out the significance of shifting political and economic relations between nation states, citizens and private sector interests. O'Brien *et al.* (2009) show how the dominant global trend in liberalisation has generated new forms of vulnerability to climate change (and wider losses in human wellbeing) in Norway and New Zealand, where comprehensive public welfare provision has been retrenched. This same research also used social contract theory to help identify social groups that are currently marginalised from national political decision-making yet nonetheless impacted by it. This included Pacific Islanders in New Zealand displaced by climate change, food security for Inuit communities in the Canadian Arctic and in Norway responsibility for future generations whose wellbeing is ever more closely linked to the legacy of Norway's oil economy of today.

Given its range of application it is not surprising that criticisms have been levelled at the social contract theory. Communitarians challenge its atomistic notion of humans; feminists and post-colonial scholars argue against its propensity to exclude; postmodernists dislike its reliance on abstract universals rather than the situated here and now (Hudson, 2003); and from resiliency theory comes the warning that climate change cannot be analysed at any one scale alone (O'Brien *et al.*, 2009). Using social contract theory to frame studies of adaptation to climate change requires rights and responsibilities that can be clearly defined, and this is not always the case particularly for future generations and distant populations connected to local events and decisions through the globalised economy.

For climate change adaptation, the most important critique of social contract theory is arguably the difficulty with which current work can move from a state to a multi-scaled/multi-actor analytical perspective; one where dominant norms and their reinforcing institutions are coproduced at all scales from the global through national to the individual and where power cuts across and works inside national administrative boundaries. In this sense social contract theory benefits from working alongside risk society as part of an analytical frame. If progressive adaptation is to address root causes as part of transformational adaptations then this is an essential area for theoretical and empirical research.

There has been some movement to extend social contract theory to the global political-economy and acknowledge global capital as the dominant centre of power through global corporate social responsibility as a contract between private sector and the consumer/producer (Zadeck, 2006). But more problematic are those growing cases where lines of influence and associated responsibility are made increasingly indirect under economic globalisation (White, 2007). Here the social contract can offer a starting point and help characterise the nature of interrupted or unclear responsibility. But global consumers, unlike national citizens, have little capacity for attributing responsibility, ascribing legitimacy and retrieving power. Here dominant power is footloose, beyond the direct control of individual nation-states. The globalisation of civil society and collaboration between governments to regulate business at the regional or global scale provide some scope for action, but so far with limited effect (White, 2007).

Human security

The social contract and risk society allow us to see adaptation to climate change as embedded within ongoing development struggles for rights and power. Human security provides a closer lens on our specific domain of interest: the play-offs to be made in balancing rights and risks between actors, and over space and time in the shaping of security. That human security is a product of the underlying cultural assumptions of risk society and institutional rules symbolised by the social contract is neatly indicated by the definition of human security used by the Global Environmental Change and Human Security (GECHS) programme. This group has a special interest in the interactions of human security and global environmental change; they define human security as:

> a state that is achieved when and where individuals and communities have the options necessary to end, mitigate or adapt to threats to their human, environmental and social rights; have the capacity and freedom to exercise these options; and actively participate in pursuing these options. (GECHS, 2009)

Human security is then a counterpoint to national security as an objective for adaptation surrounding catastrophic events and climate change more generally. National and human security can be reinforcing, but as Ken Booth (1991) argues, states cannot be counted on to prioritise the security of their citizens. Some states maintain at least minimal levels of security for citizens to promote regime legitimacy, but are unmotivated to go further; others are financially or institutionally incapable of providing even minimal standards; while still others are more than willing to subject entire sectors of society to high levels of insecurity for the economic and political benefit of others who then use their power to support the regime. Adaptation to climate change will be framed by such contexts and can offer both policy justification and practical vehicles for promoting the status quo, regressive or progressive change in human security and the rights and responsibilities in the social contract that it is built upon.

The UNDP's 1993 and 1994 Human Development Reports advanced human security as a person-centred rather than state- or even region-based approach to security. It presented a holistic and global version of human security as security from physical violence; security of income, food, health, environment, community/identity; and political freedoms (Gasper, 2005). According to Pinar Bilgin (2003) the UNDP's position was that the concept of security should be changed in two fundamental ways: (1) the stress put on territorial security should be shifted towards people's security, and (2) security should be sought not through armaments but through sustainable development. In 2003, the Commission on Human Security (CHS) developed this agenda through a basic needs approach. According to the senior researcher on the project: 'the goal of human security is not expansion of all capabilities in an open-ended fashion, but rather the provision of vital capabilities to all persons equally' (Alkire, 2003: 36). Gasper

(2005) asserts that while the concept of human security elaborated by the CHS is essentially a widely conceived yet prioritised arrangement of basic needs fulfilment, the discourse that arises from the policy framework is embedded in the concept of human rights. He argues that the integration of the two previously disparate frameworks is the critical contribution of the human security perspective, with each framework supporting the other. The basic needs model is supported and enhanced by its association with a human rights framework as much of needs-based planning has in the past adopted money-metric approaches to aggregate across people (which can lead to perverse, unintended outcomes) whereas from a human rights perspective, no individual can be sacrificed (Gasper, 2005). Moreover, unlike basic needs approaches that focus on specific claims for and by the needy, human rights discourse and practice is geared towards generating duties and seeking accountability through legal structures. Conversely, the human rights framework is supported and enhanced by its association with a basic needs model. Drawing from the work of Johan Galtung (1994), Gasper reminds us that a human rights framework tends to direct our attention towards individuals rather than structures.

The drawing down of analytical and policy lenses from the state to individual through human security complements well the observed need for social contracts to work beyond the state as nation-states become arguably less powerful than the globalising international superstructure populated by private sector and civil society interests and unelected inter-governmental or super-national bodies. Many perceive the emerging global institutional structure of governance to be as potentially threatening (or as potentially unresponsive) as the states it has so recently marginalised. As Duffield (2007) points out, the consolidation of supranational administrative bodies has not subsumed the power of metropolitan states, but rather aligned them alongside supranational powers in contraposition to the weak, underdeveloped and thus potentially dangerous states of the political periphery. From the perspective of global powers, the major threat to the security of the North is not from aggressive states, but from failed ones (Hoffmann, 2006). Thus the stage is set for unprecedented amounts of North–South interventions.

Nevertheless, it is the depth of these interventions rather than the number that is worrying to some analysts. There is an assumption that new forms of governance are no longer primarily concerned with the disciplining of individual subjects as docile citizens of particular states (though that continues) but now are combined with unprecedented levels of coordination and penetration (from supranational organisations to village committees) to produce desperation-free zones, thereby diminishing the threat of the South to the North. Human security is one of those frameworks. This is a serious warning, but in bringing together needs and rights approaches human security has the potential to bridge the public–private dichotomy that under the global liberal consensus has seen a marginalising of the state in favour of private actors. The importance of regulating private behaviour and the need to build capacity in local and national government is supported by human security but held in constructive tension with the rights of individuals. As with the social contract, context, history and the viewpoint of those at risk are arguably the most significant features in judging legitimacy and determining whose security is being prioritised and at what cost through adaptation.

Disasters as tipping points for transformation

We have seen through risk society the dangers of a naturalised modern worldview operating across all scales. The social contract has described the distribution of rights and responsibilities in society and shown this to be held in place through a balance of legitimacy and power. Human security adds to this an understanding that the rights and basic needs of individuals do not always coincide with those of the state and that play-offs in rights and risk are part of everyday development. In this discussion it has also been asserted that disaster events associated with climate change related hazards provide a distinct moment of challenge to established values and organisational forms that embody power relations. This section reviews existing secondary evidence to support this assertion as a precursor to detailed case study analysis in Chapter 8. The aim is to establish the extent to which disaster events provide leverage for academic study and also for practical movement that might be described as transformational. This should not be seen as suggesting that developmental periods outside of disaster are any less important, but simply that their review is outside the scope of this book.

The literature reveals that scholars and practitioners have long observed that the socio-political and cultural dynamics put into motion at the time of catastrophic natural disasters create the conditions for potential social (Carr, 1932) and political (Pelling and Dill, 2010) change, sometimes at the hands of a discontented civil society (Cuny, 1983). Pelling and Dill (2006) review a number of studies showing a government's incapacity or a lack of political will to respond quickly and adequately to a disaster representing a break in the social contract, while simultaneously revealing a provocative (albeit temporary) absence of instrumental state power. The destruction/production dynamic triggered by disaster creates, temporarily, a window of opportunity for both novel and traditional socio-political action at local, national, international and now supranational levels. This interpretation does not derive from an environmental determinism: it is not claimed that disasters cause socio-political change but rather that the instability generated by development failures made manifest at the point of disaster open scope for change. Indeed over the long-term there is ample evidence that human societies survive dramatic shifts in environmental conditions through a range of culturally specific adaptations (Rapparport, 1967; Waddell, 1975; Torry, 1978; Zaman, 1994). Hidden within this, though, are moments of short-term disruption, with the potential for long-term consequences.

Political change has been most comprehensively studied from drought events and related food insecurity crises (Glantz, 1976). These tend to unfold slowly and consequently are more clearly a product of development failures than rapid onset events which continue to be conceptualised as outside of human responsibility. A clear example of the interaction of environmental and political change comes from Ethiopia. In 1974, Emperor Haile Selassie of Ethiopia was ousted by a Marxist insurgency led by General Mengistu, who in turn oversaw his own government collapse in 1991. Both regimes were destabilised because their leaders failed to adequately address the deepening and progressive spread of drought, which in both cases originated in the drought-vulnerable northeast but moved southwards to envelope vast regions of the country in famine and social unrest (Keller, 1992; Comenetz and Caviedes, 2002). Violent conflict, blockades and the purposeful rerouting of supplies for political reasons have been identified as triggers in drought associated famine. Even when food is not used as a weapon, delays and mismanagement in the early stages of drought make it increasingly difficult to mitigate the full social impact (Sen and Dreze, 1999). Moreover, researchers have shown that international aid has in some cases exacerbated rather than alleviated suffering (de Waal, 1997).

To add some breadth to this discussion, Table 5.2 summarises a number of nationally significant rapid-onset disaster events and their political outcomes. It includes geophysical alongside hydrometeorological hazard contexts to demonstrate the range of interactions between political change and disaster. Cases are organised according to the political context of the polity in question: post-colonial security: modernising nation-states; Cold War security: political stability; international economic security: liberalization; and global security: advanced privatisation. Each period describes the overarching political contexts and source of pressure. Tensions in early contexts are dominated by ideological competition between state and counter-state ideologies from neo-colonial control, to nation building, proxy tensions sponsored during the Cold war; more contemporary contexts include greater influence from organised non-state actors in national and international civil society and the private sector but also a return to international political influence.

In Table 5.2, there are examples of regime change opening democratic space following disaster (East Pakistan/Bangladesh, Nicaragua, Mexico), but also cases where neo-colonial or national autocratic powers tighten their hold on the national policy (Puerto Rico, Dominican Republic, Haiti). Elsewhere disasters serve as political capital in ongoing competitions within the political elite (China) and between competing ideologies (Guatemala) including those with armed struggles (Nicaragua). The most recent events show the complexity of civil society-state relations with civil society demonstrating both regressive and progressive impetus for change (Turkey, India), the influence of international civil society and intergovernmental actors (Sri Lanka). However, even where civil society is strong and organised the power of dominant political discourses to maintain the status quo and provide opportunities for exploitive capital accumulation in the face of development failure is impressive (USA).

Comparative analyses of disaster politics show political change is most likely where disaster losses are high, the impacted regime is repressive and income inequality and levels of national development are low (Drury and Olson, 1998). Albala-Bertrand (1993) also observes that the political, technological, social or economic effects of disasters are explained primarily by a society's pre-disaster conditions, and a government that immediately marshals what material and discursive powers it has may be rewarded with improved levels of popular post-disaster legitimacy regardless of culpability. This final point emphasises the depth of the cultural underpinnings in the social contract, recognised by both Gramsci and Habermas, that can allow discourses to be manipulated by those in power. Work by Pelling and Dill (2006) confirms this analysis but also shows that competing discourses can establish a critique when building on pre-disaster political momentum.

Conclusion

The aim of this chapter has been to make a claim for transformation as a legitimate element of adaptation theory and practice. In doing so the challenges of escaping the fragmentation of modernity (Beck, 1992), the alienating loss of power to the global (Castells, 1997) and need to re-assert human rights and basic needs in an increasingly unequal world (Gasper, 2005) have been revealed as arguably the most fundamental challenges facing development and the social relations that underpin capacities to adapt to climate change risk.

The extent to which adaptation to climate change can embrace transformation will depend on the framing of the climate change problem. Where vulnerability is attributed to proximate causes of unsafe buildings, inappropriate land use and fragile demographics adaptation will be framed as a local concern. This is more amenable to resilience and transitional forms of adaptation. However, if vulnerability is framed as an outcome of wider social processes shaping how people see themselves and others, their relationship with the environment and role in political processes, then adaptation becomes a much broader problem. It is here that transformation becomes relevant.

How vulnerability and adaptation are framed have clear implications for apportioning blame and the locus of adaptation and its costs. Where vulnerability is an outcome of local context then it is local actors at risk who will likely carry the costs of adapting (for example, through transactions and opportunity costs incurred through changing livelihood practices). Where vulnerability is seen as an outcome of wider social causes then responsibility for change becomes broader, possibly more diffuse and less easy to manage and certainly more likely to touch those in power. These two approaches to the framing of vulnerability and subsequent adaptation are akin to the distinction between treating the symptoms and causes of illness.

Transformation does not come without its own risks, inherent in any project of change is uncertainty. History is replete with examples of transformation social change being captured by vested interested or new elites. As noted with regard to human security, both the poor and powerful are aware of the costs of change and prefer the known even if it is a generator of risk. As climate change proceeds and mitigation policy fails the potential for dangerous climate change increases. This forces us to reappraise the potential costs of transformation set against business as usual. Handmer and Dovers (1996) warned against the sudden collapse of

Table 5.2 Disasters as catalysts for political change

					Post-colonial security: modernising nation-states			
Affected city/ country	Year	Pre-disaster state civil society relations	Hazard and loss	Local/regional government/civil society response	National government response	International response	Socio-political impact/change/legacy	
Puerto Rico	1899	Recent political independence from Spain; rising US economic interests; emergent labour unions; local municipalities key political actors	Hurricane San Ciriaco; 28 days of rain; huge crop damage; estimated 3,100–3,400 killed	Municipalities: distributed relief; assessed damage proposed financial plan for recovery	Cooperates with acting US military governor to receive aid; popular resentment and finally acquiescence	US uses humanitarian aid to undermine nascent nationalism movement and to solidify national influence	State 'Anglicised'; social hierarchy adapted to economic modernisation; elites funded; 'deserving' poor become workforce; union gains reversed; cross-c ass bid for independence thwarted	
Dominican Republic	1930	US-groomed dictator Trujillo in power; coopting of civil society and suppression of political opposition	Hurricane Zenon destroys most of capital city; estimated 2–8,000 killed	Unknown	Immediate request for international aid; reconstruction funds used to build city as symbol of a modern nation-state and presidential power; renamed Ciudad Trujillo	US supports regime and reconstruction	Entrenchment of new dictatorial (right-wing populist) regime; nation–state modernisation continues with ethnic cleansing of Haitian labour migrants	
East Pakistan (Bangladesh)	1970	Deep political tensions between East and West Pakistan with economic and ethnic underpinnings; an organised independence movement in East Pakistan	Cyclone Bhola estimated to lead to 500,000 deaths	Local government overwhelmed; with no help from central government in West Pakistan, citizen support of local leadership swells	No disaster plan; state paralysis; post-election political repression	Chaotic: US arms (West) Pakistan; humanitarian efforts directed to respond to massive refugee crisis in India	Complex political emergency; East Pakistan leadership declares independence; Bangladesh established as state in 1971	

Cold war security: from modernisation to political stability

Affected city/ country	Year	State/civil society	Hazard and loss	Local/regional government/civil society response	National government response	International response	Socio-political impact/change/legacy
Haiti	1954	Predatory state; landed peasantry; relative openness (labour unions); major infrastructure modernisations	Hurricane Hazel destroys cash crops; estimated 1,000 killed	Unknown	No disaster plan; corruption soars with international aid	International funds flow; Catholic Relief and CARE begin first work in country	Regime corruption sparks cross-class protests; US-trained military takes control; Papa Doc Duvalier cuts deal with military leading to a long lasting and entrenched violent dictatorship
Managua, Nicaragua	1972	Dynastic dictatorship; civil society repressed; elites disenfranchised; vocal opposition movement	Earthquake destroys much of capital city; estimated 10,000 killed	Extended families provide relief; the city is evacuated	No disaster plan; focus on physical reconstruction of capital and repression of civil society	International funds flow; gross corruption by elite; military appropriates development	Corruption provokes anger; liberation theology and Sandinismo provide oppositional discourse; social capital developed during recovery period feeds into cross-class revolutionary movement leading to regime change
Guatemala	1976	20 years post CIA-coup; a military state; technocratic president; slight opening for human development; active opposition	Earthquake destroys parts of capital and villages of central and northern highlands; estimated 23,000 killed	Municipalities inadequately funded; peasant groups and Church respond	No disaster plan; focus on physical reconstruction of capital and repression of any non-state organised activities	International assessment teams remain only in capital; few foreigners have firsthand knowledge of high losses in rural indigenous villages	Military threatened by post-disaster peasant organisation in context of active insurgency; state represses indigenous earthquake reconstruction projects; guerrillas use earthquake as oppositional discourse (time for change) for organising purposes; counter-insurgency escalates; insurgency escalates

Table 5.2 (continued)

				International economic security: liberalisation			
Affected city/country	Year	State/civil society	Hazard and loss	Local/regional government/civil society response	National government response	International response	Socio-political impact/change/legacy
Tangshan, China	1976	No theoretical distinction between state/civil society; a period of political transition during the last days of Mao and the cultural revolution	Tangshan earthquake destroys important industrial city; estimated up to 655,000 killed	Massive self-help campaign; city requests and receives funds and relief from regional administrations	Nationally significant disaster plan (prediction) fails; reconstruction distorted by massive political struggle between Maoists and reformer Hua Guofeng	International aid refused; West denied access and information	Earthquake appropriated as political symbol for loss of 'Mandate from Heaven' (oppositional discourse); Cultural Revolution ended; return to previous plan for modernisation and liberalisation of economy
Turkey	1999	Authoritarian secular state, democracy, strong religious civil society seen as threatening secular state traditions	Marmara earthquake; estimated 17,000 killed	Limited, failure to regulate construction a major cause of loss	State slow to respond, local and national civil society (religious and secular) filled vacuum	Constrained by state failures to coordinate response	Civil society demonstrated capacity to provide social support; state responded by closing bank accounts of religious groups in particular
Gujarat, India	2001	Democratic system with a strong civil society; hierarchical	Earthquake in Kutch district; estimated over 20,000 killed	Limited in contrast to widespread civil society mobilisation	Initial response slow, ad hoc and chaotic	Widespread; support for participatory reconstruction schemes from multilaterals	Response reinforced the strength of civil society in India; reconstruction criticised for exacerbating socio-cultural inequalities; some associate this with subsequent religious riots in Gujarat in 2002

Global security: advanced privatisation of national economies and services

Affected city/ country	Year	State/civil society	Hazard and loss	Local/regional government/civil society response	National government response	International response	Socio-political impact/change/legacy
Nicaragua	1998	Electoral democracy; free press; active civil society politically polarised	Hurricane Mitch; 2,000 die when entire town buried in mudslide; estimated 3,800 killed in total	Following a decade of state downsizing, civil defence, fire and police poorly staffed, resourced, and disconnected from central government and scientists, unable to function properly	Government scientists report on impending storm; President denies crisis	World Bank and UNDP sponsor the development of a national disaster reduction system; international mediation to open governance in reconstruction	Brief opening of discourse between state and civil society development actors under international mediation; joint development of a reconstruction plan; retrenchment and re-imposition of pre-disaster political culture with strengthened leverage for global economic interests; a lost opportunity for social reform
Morocco	2004	Authoritarian kingdom; failure of political liberalisation; civil society weak	Earthquake strikes marginalised region; kills more than 560	Concentrates aid in port town; refuses to extend appropriate aid to villages	Spends almost equal amounts of aid monies on reconstruction and repression	US and European countries compete to come to Morocco's aid	First political mobilisation in Riff mountain region for many years; youths protest failure of state response; neoliberal political economy; state repression

Table 5.2 (continued)

Affected city/ country	Year	State/civil society	Hazard and loss	Local/regional government/civil society response	National government response	International response	Socio-political impact/change/legacy
Sri Lanka	2004	Entrenched political and armed conflict between Sinhalese majority and Tamil minority; Muslim minority marginalised from both; an electoral democracy with limited but free press	Tsunami devastates ⅔ coastline; 35,322 killed	Civil society in rebel held areas especially prepared for emergency response	Concentrates resources in government-held and economically important regions	Massive supranational and international humanitarian and geopolitical response; swamps local capacity and reignites political tensions	International interventions fail to support transition from ceasefire to peace accords; fishermen worst affected sector; many barred from returning to home site while hotels acquire land; civil society continues to operate in a war zone
New Orleans, USA	2005	Electoral democracy; free press; strong civil society and private sector interests; voter alienation	Hurricane Katrina floods city and region: 1,836 confirmed dead; more than 700 in New Orleans	Mayor does not want to alienate business leaders by calling for mandatory evacuation; acts too late; governor fails to convey urgency of needs	Federal government fails to act on warnings that levees might be breached	Some international aid accepted but also politicised, e.g. offers of aid from Venezuela and Cuba	Nation undergoes intense but brief analysis race/class relations; impact of neoliberal policies on disaster reduction now under scrutiny; maladaptive development under scrutiny; real estate speculation and investment soars in flooded region

social systems that respond to threats with only limited, transitional change. The prospect that without transformational adaptation undertaken with some measure of planning and inclusivity dangerous climate change may force uncontrolled and more anarchic forms of transformation onto societies is worthy of consideration.

Part III

Living with climate change

6   Adaptation within organisations

What matters is not structures, but relationships
                                    Scientific Advisor to the Welsh Assembly

This comment, made by a scientific advisor to the Welsh Assembly, is a very clear acknowledgement of challenges facing managers having to consider the organisational challenges of climate change risk management alongside existing imperatives including efficiency and transparency. Here our respondent was clear that while formal institutional structures are necessary to give organisations shape and direction, when adaptation is required to protect core functions this is nuanced – potentially championed – by the contingent, shadow world of informal relationships. This chapter presents the viewpoints of actors within two different kinds of organisation who reflect on the interplay of social relations within canonical and shadow systems that characterise adaptive capacity. Communities of practice and networks of looser ties are considered. The aim is not simply to illustrate adaptation as resilience but rather to give some substance to the complexity of social relations that give rise to adaptive capacity originating from within organisations. As noted in Chapter 3, while resilience may be the dominant external outcome of the social agency described within organisations, internal acts that could be classified as transitional and arguably transformational are also observed.

The empirical evidence presented draws from interviews held with members of the UK Environment Agency active in Wales, and a dairy farmers' cooperative from Carmarthenshire called Grasshoppers. Earlier work (Pelling *et al.*, 2007) has provided a synthesis of these interviews and also with those from scientific advisor groups to the Welsh Assembly. The aim in this chapter is to provide a detailed examination of the viewpoints of key informants reflecting on their relationships with organisational structures and other actors to use or open space for social learning and self-organisation. Such internal acts of adaptation targeting institutional modification are identified, as are adaptations directed at the external environment.

The following section provides policy and methodological context for the empirical data, which is then presented.

Context: policy and methods

In the UK, statements by DFID (2004b), GNAW (2001) and MAFF (2000) have highlighted the dual role of public sector agencies needing both to adapt their own goals and practices to take account of climate change, whilst also shaping the enabling environment to support the adaptive capacity of private, public and civil sector actors and individuals operating within their spheres of influence. In this way it is doubly important to understand the ways in which the capacity and direction of adaptation within such organisations is shaped. Despite this only little thought has gone into planning how adaptive capacity to climate change and variability might be built as a policy imperative alongside efficiency, transparency, accountability, legitimacy and equity. Most work to date has been undertaken within the adaptive management school and there are parallels with the analysis presented here (see Chapter 2). This is important because existing bases for organising and implementing policy are challenged by the complex, dynamic, 'trans-scientific' (Weinberg, 1972) cross-epistemic problems associated with climate change. In responding there is a need to develop organisational capabilities that reflect the uncertain nature of knowledge. Central to this task is a better under-standing of the ways in which organisations learn and adapt. This is especially so when adaptive innovations challenge dominant ways of thinking and defining goals and responsibilities.

As Chapter 3 demonstrates, research on learning and adaptation to climate change has focused primarily on the influence of formal institutions and on reactive adaptation. Empirical work has shown that adaptation can be a source of contestation for political actors operating across hierarchies of scale (Iwanciw, 2004), and with contrasting ideologies; for example, with tensions emerging through the interplay of top-down command and control risk management and local self-organised adaptation (Tompkins, 2005). From the viewpoint of pro-active adaptation, Grothmann and Patt (2005) acknowledge the importance of psychological factors in determining the adaptive capacity of individuals.

This chapter presents evidence for adaptive capacity as arising out of cognitive processes (ongoing social learning) embedded in the social relationships of organisations (which are given shape by both formal and informal institutions and their practices). Such generic socio-cognitive attributes of organisations can contribute to the building of robust adaptation, responding not only to surprises associated with climate change but also the uncertainties of future economic, social and political change (Schneider, 2004; Willows and Connell, 2003). However, research in crisis management has pointed to the difficulties that can be associated with these characteristics. Organisational culture, communication practices and decision-making processes generate the conditions in which crisis events occur (Reason, 1990a, 1990b, 1997; Smith, 1990, 1995; Turner, 1976, 1978). At the same time, this research has sought to push the boundaries of contingency planning by encouraging managers to start 'thinking the unthinkable' (Smith, 2004) as a means of considering the range of problems that can arise and how organisations might be structured to anticipate such risks. Preparing organisations for the unimaginable as well as planning for the unexpected is enhanced where there are diverse social relationships with open informal space beyond corporate control. These spaces allow individuals or sub-groups within organisations to experiment, copy, communicate, learn and reflect on their actions.

Perhaps one reason for the limited literature on adaptation within organisations (compared with research on adaptation within local communities for example), and in particular on the ways in which social agency and institutions interact, is the difficulty of surfacing respondent viewpoints. Much of the experience of social learning and self-organisation happens as part of the routine practice of working within an organisation with the distinctions between canonical and shadow spaces often blurred. Elsewhere working in the shadow system is on the fringes of professional good practice and seldom disclosed publicly. The approach taken to generate the data presented below was to engage respondents in a three-stage conversation. First, respondents from each organisation were self-selecting, having responded to an open invitation to attend a workshop framed as an opportunity to reflect on the organisation's adaptive capacity and potential future strategy. Second, workshop discussions were followed up with individual interviews, or in some cases researchers were invited to follow-on meetings. Finally, summary data and analysis that had been made anonymous were circulated amongst respondents for comment and as a verification tool. The initial selection of organisations was based on existing contacts and a desire to engage with respondents working in different organisational forms with responsibility for setting the policy or information environment for other actors and businesses.

In the framing workshops respondents were presented with a low probability, high-impact climate change scenario for which no contingency planning existed in the organisations under study. The UK scenario was for strong warming over 20 years to reach a climate similar to that of contemporary southern France, followed by a collapse of the north Atlantic thermohaline circulation systems and a rapid cooling over a subsequent 10 years to reach a new climatic equilibrium close to that of southern Norway. To generate concrete examples of the role of social relations in adaptation respondents were also asked to identify past analogues for the climate change scenario. The analogues chosen by respondents differed, but common examples of external surprises were the foot-and-mouth outbreak in 2003, ongoing changes to European Common Agricultural Policy and the European Waters Directive: stressors which the organisations acted to mediate and were felt to be wide-ranging and, to varying degrees, unpredictable in their ramifications for respondents' organisations.

The range of climate change impacts considered in one workshop are presented in Table 6.1. The recognition that not only were climate futures uncertain but the development impacts of any one climate future multifaceted and potentially reinforcing was key in justifying the focus of discussion on the relevance of generic, fundamental adaptive capacities built on social learning and self-organisation rather than a search for material adaptation policies.

Table 6.1 Warming and cooling scenarios for Wales

	Warming scenario	Cooling scenario
Weather regime	Increased winter rainfall and flooding Higher temperatures overall Hotter, drier summers A similar climate to that of Bordeaux	Increased flooding in spring due to snow melt Lower temperatures overall Colder winters with one in seven winters 'extreme' A similar climate to Oslo
Rural development	Diversified economic opportunities Increased rural population	New opportunities for secondary employment Rural depopulation Transport disruption and less accessibility to services during winter
Public health	Increased respiratory disease in wetter winters New diseases Heat stress Pollution effects?	Increased respiratory disease in colder winters
Agriculture	Soil loss due to flooding New pests and diseases Late summer grazing reduced but may be compensated by increased grass production overall More difficult to use land effectively Crop diversification possible, especially on the coasts, but soil quality may limit this	Soil loss due to flooding Reduction in stock or capital spending on winter housing Loss of winter growing season – less grazing implies less protein production
Forestry	Timber productivity up while quality down Use of trees for water management?	Timber productivity down, while quality up Pressure on forestry management More forestry on marginal rural land?
Biodiversity	Links between habitats forming wildlife corridors gain importance to allow species migration More active management of species migration needed under warming than cooling scenario Loss of key species like sphagnum moss Pollution effects?	Links between habitats forming wildlife corridors gain importance to allow species migration Eco-restoration possible as climate cools from a preceding high?

Table 6.1 (continued)

	Warming scenario	Cooling scenario

Tourism	Higher volumes anticipated
	No extended winter slow season
	Improves in comparison to
	  competitor destinations
	Storm and flood risk to
	  infrastructure
	Loss of 'Green Hills' image
		Lower volumes anticipated
		Possibility to develop winter
		  sports
		Seaside market in decline

Other
industries
	Less vulnerable water supplies
	  than in England but may be
	  indirectly impacted by
	  English extraction
		Shellfish production crashes
		Possible loss of high-tech and
		  footloose industries

Note: Additional empirical analysis is available on the project website, http://rcc.rures.net

Case study analysis

The aim of this section is to reveal the interplay between institutions and individual action that construct the relational space for adaptation within organisations. The dominant form of adaptation considered is resilience. The two organisations included in the discussion allow two different sides of adaptive capacity to be examined. First, in the Environment Agency, responsibilities for setting the operating environment for more local organisations to adapt are explored. Second we use efforts of a farmers' support group to facilitate aspects of adaptation for individual farmers. In both cases the assessment of capacity to adapt to climate change is forward looking. That is, we do not seek to describe assets used in past rounds of adapting to climate change. Rather we explore the social relationships and actor behaviour that constitute these organisations as a way of mapping out capacity for adaptation based on the theoretical arguments made in the preceding chapters. This frees analysis of capacity to adapt to climate change from a historical determinism which would skew and limit results where future events associated with climate change may be very different from past experience. In both cases the aims of the organisations are to promote adaptation as resilience. There are though examples of individual actors attempting to change the direction of the organisation; this is especially so for the Environment Agency. These serve to exemplify the skills and strategies that can enable transitional adaptation within an organisation.

The discussion for each organisation is presented around a series of quotations. This gives voice to the respondents but also provides a contextual richness that would be lost if a summary alone was provided. Themes of social learning and self-organisation help to structure the accounts. Self-organisation is unpacked further by statements on the interplay of shadow and canonical systems and of social communities and networks acting within and across the organisations. Data emerged inductively and act to verify these attributes of adaptation that have so far been described largely in theory. Respondents and in some cases secondary organisations are not named to maintain confidentiality.

The Environment Agency

The Environment Agency is a key mediator for climate change adaptation in the rural sector in the UK. It is charged with protecting and improving the environment and promoting sustainable development including flood risk management in England and Wales. It acts both to regulate and advise on rural development.

Respondents discussed capacity to adapt to possible future impacts of climate change through focusing on their personal and professional experience of constraints in the canonical system, the role of the shadow system and how together they form an institutional architecture for adaptation. Many of the observations are not tied directly to experience of climate change associated events or policy but speak to the generic interaction between professionals and institutional structures within the organisation. The uncertainties that climate change brings and the knowledge that past events are increasingly inappropriate as guides to future crises makes such knowledge central to understanding and potentially supporting adaptation to climate. What follows is not an assessment of adaptive capacity across the Environment Agency but rather a reporting of viewpoints from key informants working as professional scientists from different points within the organisation.

Institutional constraints

Taking or designing adaptive actions is facilitated or constrained by existing institutions, which have their own logic, history and transactions costs if being reformed or dismantled. Thus an important type of observed proactive adaptation was institutional modification: efforts to reduce conflict between adaptive possibilities and existing social realities, and so create enhanced opportunities for adaptive actions to arise as needed. The impetus for this can come from without or within the policy system, for example:

> In a sense, we're doing that [institutional modification] through our seminars, but we are also working in the Welsh Assembly and the Environment Agency, and everybody else. We're trying to get the Welsh Assembly to lead on a Welsh climate change communications strategy. It's not a priority for them, but we are trying to lobby for that.

Institutions affecting adaptive capacity and action were found to have a fluid quality. They were renegotiated as circumstances changed, as different individual and organisational actors became involved and as existing actors readjusted their internal priorities. For example:

It is set in their contract that they have to do a workshop and that it needs to have these outputs, but there is nothing in it that says you have to do it in this way. But if one of us were to say to someone, look we think you ought to do it this way, then they're not going to say no. They might come back and say that they've had a better idea.

Negotiation is an asset for facing the uncertainty induced by climate change. But this has financial and other costs. Considering how institutions do or might change necessitates an analysis of the power configurations that conserve or act against particular institutions. Power relations can be given expression in many different ways, but in an organisational context, the direction of resources is an important one. As the respondent notes, though, agent led external action is challenging:

> Politics is difficult. I have certainly tried to foster close relations with DEFRA, DOE, DETR whatever it happens to be, but you are dealing with a culture that is fairly rigid there – they pay the bills, we do what they say.

Institutions can both constrain and enable adaptation. For individuals seeking to influence organisational behaviour and direction this revealed a tension between personal and/or professional agendas. This was particularly difficult when it felt as though institutions originated hierarchically, and the costs of renegotiation were exorbitant for the individual:

> In the day job there is a day job. I have objectives to do. What I do outside of that is my affair so corporately the culture is quite thick – quite hierarchical, which is frustrating because if we are moving from managing simplicity in regulated resources through to managing complexity – environmental systems – one of the first tenets is devolution of decision making and yet we are going diametrically the opposite way so I find it frustrating intellectually certainly personally.

Social learning is central to adaptive capacity. It can be indicated by changes in capacity to act arising through experience – for example, through institutional modification creating an atmosphere where learning is promoted is part of the shaping of adaptive capacity – and can be the difference between important experiences being overlooked, forgotten or translated into enhanced capacity to deal with future climate-change-related uncertainty and threats:

> There clearly has been a lot of learning: Enquiries etc., and people presenting information back to us. It's had a big impact on how we organize ourselves. It's created new areas of work and funding to tackle gaps. . . . The lessons are quite general and cross-cutting: How do you get bad news up the line quite quickly? How do you ramp up resources quickly? . . . That can now happen very quickly. Not only are there plans to show us how to do that, but we have practice simulations.

Opportunities for learning arise throughout organisational life, and can be fostered: 'We have informal lunchtime sessions, and people ask questions about it. The questions will be more informal. People are sitting there eating lunch and asking questions. It's informal in that respect.'

On the other hand, not all learning is positive. One respondent warned about uncritically accepting the lessons of past experience, without continuing to probe their relevance to new situations; a key lesson for climate change adaptation, but one that is difficult to institutionalise:

> I think one issue that is quite difficult is learning from experience. One has to be very careful that the experience you had is relevant to the problem that you now have. We often come up against the situation where people who've had long experience say 'Oh yeah, we tried that, and it didn't work. That's it.' It cuts off the options and one has to very careful that one is saying that was the experience, but was the context and the problem the same?

Good communication skills are a necessity for institutional modification, something that a number of interviewees demonstrated, including strategies for formalising and adding value to knowledge through external collaboration. This was a particularly effective – but time consuming – method for influencing higher up the hierarchy or across sectoral and professional barriers. Relevant for slow onset and long-term adaptation measures this strategy for crossing the internal barriers within organisations is too slow to respond to rapid and extreme events:

> That is why I write so much. If it is out there in the white literature then it is in the public domain. A peer review paper has more weight than my opinion – particularly when I bring in co-authors who happen to be lawyers.

Successful communicators had cultivated linkages across different epistemic communities and saw themselves as conduits of information and points of influence shaping spaces of adaptive capacity within and between both communities and their representative organisations:

> The XXX, which is a national organization . . . has done a tremendous amount and in some instances the Agency is being perceived as an obstacle and in some ways it is being perceived as an ally, but there is a risk of that relationship being lost and because I am on the board of various other charities and I'm giving a key note at the XXX meeting on Tuesday. I've got a very direct personal relationship there and I'm publishing papers in my own name, not using work time whatever to get the learning from that, put it in the right literature so I can go to the policy people in the Agency to say LEARN, you don't have to trawl through grey literature, unpublished sources here is all the right literature put together – APPLY IT, DO IT please. So yes, I'm keeping doors open, but that is a personal mission and I don't expect that will be a particularly common occurrence throughout the organization.

Learning with wider stakeholders, and especially the public had its costs with a difficult balancing act between efficiency and building adaptive capacity; for example, by protecting staff so they might undertake their work without too much interruption from other stakeholders. The following comments respond to a recently established telephone call centre:

> In terms of the general public what is happening corporately is walls are being built so I think we are going in the wrong direction. You know if you are re-engineering an organization where your front line, your regional and area staff are delivery merchants then you want to stop then 'wasting time' in dialogue with the punters. You want them to be doing stuff, not talking about stuff.
>
> . . . a lot of the public trust that the Agency does engender, it does not engender a lot but, a lot of that is simply because the local officers know the local people and the local issues. So actually I fear that what we are doing is losing the connection. I think the call centre is going to make us become a big impersonal monster . . . It is a personal view this, I think we are losing an important part of our relationship with people . . . the personal relationship with the regulator is vital . . . That sort of delivery of service model [the call centre] is what the Agency's reorganization is about, so it is successful in those terms – but, you know, not in terms of being in touch with the environment and people who are active in the environmental sense.

Communication that can help build capacity to adapt to climate change requires skills such as knowing who to communicate with, how to find them and how to communicate effectively, and designing acts of communication which are appropriate to the task. Communication is not a neutral act, and there are many conventions that apply to the way that communication is carried out in different relationships and contexts. Because the appropriate combination of learning and communication strategies available to actors is determined by the cultural characteristics of the organisational setting in which they operate, it makes sense to speak of the knowledge culture of an organisational setting. That is the characteristics of an organisation or other social body that make particular forms of learning and communication possible or not. The sense of a pervasive way of being that both influences the individual and that results from the collective actions of individuals came through clearly in one interview:

> So to what percentage am I attributable? I don't know. To what extent is culture changing around me and these ideas becoming more and more? I don't know. I can't measure that, but in my own head I'm pretty well convinced that I have banged on at certain people for long enough that we have got an understanding.

An important aspect of adaptive capacity revealed by looking at learning and communication in terms of a knowledge culture was that the informal and the tacit are just as important for knowledge as formal and explicit channels, even from the organisation's perspective. For example, in the case of learning, formal learning was in some cases identified with training, but it was clear that this was just one aspect of learning from the individual viewpoint. Thus, throughout the interviews a range of evidence referred to informal channels of learning and communication, and the ways these were rooted in both formal and informal activities and institutions.

> So yeah formally, in the formal email, telephone whatever you play the game but you still carry out the learning stuff. If I see the head of xxx who I know very well and for many years I'll say 'Have you seen this paper?'. 'No I haven't actually.' 'Oh I've got a few on the line, have you got a minute ...?' 'I've got this one on common law', you know, 'I've got this one on economics'. 'Yeah OK, let's talk about that, that's really interesting blah blah'.

Adaptation and the shadow system

This section provides support for the claim that shadow systems are an important source of adaptive capacity. Most interviewees could identify an informal shadow system, and argued that the informal is an essential part of organisational life: 'The way I think is that the day job is largely defined by the delivery of regulation and the influencing stuff happens through the informal routes by and large.'

Shadow systems are unobserved by the canonical and allow risk taking. Adaptive management has the ability to experiment and take risks as a core tenet. The benefit to the canonical organisation of the shadow system arises through a degree of alignment between actors' formal roles and their informal skills and capacities. Thus the personal capacities of individuals to wield influence and to work with knowledge became part of the organisation's capacity to adapt: 'I know that statements I have made and discussions I've had with very senior people have later turned out in more or less verbatim in strategy documents.'

While individual initiative within the shadow system cannot be planned for, it could be incentivised, opening up a major adaptive resource for the organisation:

> The organization three years ago had a tokenistic approach to the social, but now has social policy. This is moving more and more mainstream, and arguably there is sort of a change in political direction anyway, but an individual mover and shaker who I happen to talk to quite a lot has been singularly effective in raising that as a policy.

Conversely, this allowed individuals to enact their values through the operation of the formal organisation, uncovering contrasting types of legitimate behaviour:

> That it depends who you ask these questions to. There are those who work hard to get the job done. There are other[s] who have moved between

different organizations and have some weird idea to try and change the world and migrate around the place to try and do that.

My private action has feedback into the organization.

The re-alignment of formal and informal knowledge networks in this way is an example of agent-centred resource management. This helps the organisation learn about its environment, improving adaptive capacity, even when the canonical structures build barriers to communication and flexibility:

And then we get back in our boxes and I don't communicate with him because he is not part of my section.

Management tends to perceive that [personal lobbying] as rocking the boat so I have kind of given up.

This suggests that an important area for working with adaptive capacity is positioning the role of canonical management with respect to the shadow system. This is not straightforward. The shadow system is almost by definition resistant to management effort. But while it is not necessarily manageable, there is scope for management activity with respect to shadow systems. The simplest strategy is perhaps to recognise the role of the informal and to accept a degree of imprecision and failure when risks are taken, allowing spare capacity in planning including providing time and flexibility for individuals to work around the formal system where required. This is not straightforward, and a key problem is providing examples of outcomes from working the shadow system where these are often indirect:

How do I demonstrate that by going to this meeting rather than that one that a particular outcome came about? It's all about influencing, but only sometimes can you point to a report or a policy document and show that they've used your wording.

Thus a more positive strategy with respect to the shadow system might be to find ways to report on it and to incentivise individuals to use their skills in creating and maintaining informal relationships for the corporate good. Above all, it is a matter of making sure that the individual skills are available in the first place. This creates a demand for individuals with competencies relevant to the shadow system. Interviewees produced a range of examples of skills they utilised in skilled informal interaction:

Learning the ways that the organization works is the only way you are ever going to be able to influence it at all because if you try to influence it from a different discourse or dialogue you just bounce off it . . .

I write books as well and ask people to tell me what is wrong about them – this is a way of roping people in. I treat publications as a way to integrate views with some clarity and common sense.

In terms of playing the corporate game, it is about knowing to put the right, copy the right, people on emails, don't jump levels over and above bosses, all the basic hierarchical things; that is the way it works formally. The way it works informally – having been around the organization for a million years and knowing all the other people that have been in the organization a million years, you know – that is what water coolers and coffee machines are for.

Institutional architecture for adaptation

Understanding organisations and the institutions that shape them is a key part of balancing canonical and shadow space and facilitating adaptive capacity. In this section we examine respondent viewpoints on institutional architectures in terms of communities and networks that cross-cut the formal organisation.

Communities comprise groups of people who share identity expressed through similar interests and common values:

> So I am not interested primarily in a community that want[s] people to play by the rules. I am interested in people who, for want of a better word – although it is a shitty old phrase – 'want to make a better world'. In other words, if someone really cares about social factors and sustainability and they have sorted out a job in an organization that can do something, I will feel sort of attracted to spend time with them. In terms of my community it is people who are looking to make the step changes.

Community boundaries do not necessarily reproduce those of the formal organisational contexts in which they occur. Thus communities tend to arise through mutual engagement rather than management fiat, and are very much of the shadow system. But although they have their own rhythm of development it is possible to give them space to grow by making time for individuals to interact. From the individual's perspective, communities can be a significant resource, opening up opportunities for action though links with others with similar interests: 'There are other trouble-makers out there that I tend to gravitate towards. My community is people often dressed as very establishment but who are basically in the organization for their own agenda.'

Shadow communities are a natural unit for adaptive action, as shared interests and similar worldviews make negotiating and endorsing plans and reactions quicker and easier.

> For example, there's a group of farmers in mid-Wales who are looking at how they can make agriculture more sustainable, looking at how to deal with flood control, with soil quality. That's like a self-motivated group of 10 farmers, acting as a community because they see particular environmental threats. You'd have to look at groups like that to get that core of adaptation.

As with any form of organisation, communities have internal differentiation, and there can be disagreement within their membership over their shared identity and boundaries. Also membership is not necessarily mutually exclusive, and communities overlap, giving a dense texture to social architecture – Wenger's (2000) constellation of communities. Because shared interest is assumed and may be beyond challenge, they can also close down opportunities for change.

A more open social form is the network. Networks arise in social life across boundaries of difference. Thus, unlike communities, common interest is not assumed, but instead is negotiated. As with communities, interviewees were able to point to examples of networks with significance for their professional lives. Networks were a site of bridging social capital, linking together organisations and communities. The encounter with different values and worldviews that occurred through networks made engagement in networks a significant opportunity for learning:

> Yes, there's a network. If you can identify where to implement different policies . . . you can identify certain people, you can see who has done this and been quite successful at it. You build a little network of people to go to. A little expert group in a sense. It's important to learn from people, rather than start off from a blank sheet all the time.

Thus networks provided opportunities to build and operate adaptive capacity:

> The [Welsh] Assembly would need to base its case for change on reasonable evidence, and that's where it works with networking. Networking with the likes of the Environment Agency in order to say 'This is a current situation', and be able to make predictions in terms of what is likely to happen.

It may be that operating as an individual in a network requires a different skillset from working within a community. With their basis in relationships between individuals, there is a danger that forcing networks into existence will result in a paper exercise or a locus of discontent. However, there is much that can be done from a management perspective to foster networks:

> When you're dealing in a cross-cutting issue, which this [adapting to rapid climate change] would be, then you have to try to pull the people together in some sort of project group. The difficulty is making sure that that happens more than in name. You can get people along to meetings, but it requires issues to be sorted, actions to be taken, so that it permeates out into additional action, with all the resource that requires.

What both networks and communities have in common is that they are founded in relationships of trust. Within communities, trust was shown to arise from shared interest:

> You tend to know certain people, certain groups, and they establish a track record of whether they can deliver or not, because you are clearly trying to find the ones who are most effective, rather than spend a lot of time saying you want this to start from grassroots sort of thing.

In a network, trust was required in order to negotiate a mutual interest, and arose through ongoing engagement. Trust can be invested in individuals and expressed in personal relationships. However, it can also arise through institutions, from the social contracts embedded in formal organisational forms. Trust was important in adaptive capacity, because it enabled social action and decreased the amount of effort involved in maintaining communities and networks. That is not to say that creating and maintaining trust does not have costs of its own:

> If you pull that lever and nothing happens, then you lose all credibility for what it is that you're doing. It makes it clear that you don't understand what you're doing and people will therefore take no notice of you. So there's a credibility issue here in actually making things work.

Grasshoppers farmers' group

The Carmarthenshire based dairy farmers' support group, Grasshoppers, has about 20 members and was established six years before our study. Its aim is to explore what became known as the New Zealand grazing system. This system differs from dominant dairy practices in the UK through a combination of conserving hay for the winter, turning cattle out earlier in the year and calving only once a year. This results in little or no spending on winter feed and reduced labour costs. Thus although less milk is produced than under a more intensive regime the profits are greater, and the farmer has more time to pursue other interests. The intention of Grasshoppers' members is to maintain their rural livelihoods and quality of life by changing farming practices: a case study in resilience.

The members of Grasshoppers are well positioned to discuss the generic attributes of organisational relations that shape adaptation. As a group they have already demonstrated an ability to adapt proactively to changing economic conditions within the dairy sector. Their current mode of practice is probably better adapted to climate warming than conventional dairy production in the UK. Nevertheless, under an extreme climate change scenario there would be substantial challenges to be faced. Exploring the proven adaptive capacity of the group offers an opportunity to explore the role of institutions and social learning in shaping organisations to support individual farmers in planned and proactive climate change adaptation.

Group activities centred around monthly, rotating farm visits. Meetings had a sharply critical tone which, over time, had developed a culture of mutual respect, trust, fostered social learning and encouraged innovation. As with the

Environment Agency, the themes of community, network, trust and exclusion arose from discussions and provide themes for understanding the production of adaptive capacity and social learning within Grasshoppers.

Seen as a community, Grasshoppers appeared to have a strong and well-developed shared identity. Grasshoppers was created intentionally with new members being recruited through invitation only, reinforcing this shared and distinct group identity. Importantly, membership did not focus directly on joint commercial activity. Members were more concerned with sharing knowledge, improving practice and mutual support in meeting the challenges of the New Zealand system than in striking business partnerships or joint commercial advocacy: 'Sharing information is really key, something I realise from these other farmer groups compared to us.' One member likened this feeling of being in a learning community to adaptation: 'Openness and sharing information is a major part of adaptation.'

Examining Grasshoppers in terms of networks highlights external relationships, and once again the focus is on information and learning. That is, through Grasshoppers members were able to manage their access to information resources. The strongest expressed links were with dairy farmers outside the UK, drawing on contacts made from a range of contexts, because: 'Overseas is best. The UK is too mainstream [in dairy farming] – and we're not! Also there is no basic/market research in this area because there is no commercial basis so it is not picked up on.' In this case, it was clear that a wide base of information sources was a valued resource for adapting to future climate change. For Grasshoppers, this enabled both improvements in existing practices, as well as challenging adaptations, with shifts in livelihood and lifestyle goals.

> In terms of adapting to a different climate, you could go and look at places in the world where people already live with it. Now we have learnt from New Zealand, but if the climate cooled we would learn from other parts of the world.

In Grasshoppers, trust was closely tied to the duty of confidentiality, identity and membership, indicating that it arose first and foremost as a function of community building:

> Trust is very important to the group's functioning and this has taken time to build up. For example, Grasshoppers started with members sharing limited information on the purely financial aspects of the grass economy. We now share economic and other information on all aspects of farmers' livelihoods.

Trust was described as having built up over time to extend beyond members' professional affairs to finances and even friendship, the latter effectively blurring the boundaries between the canonical and shadow relationships and roles of

Grasshoppers members: 'Other than my wife and the nucleus of my family I'd talk with group members first [about a problem].' As a result, members of Grasshoppers felt they could rely on the information they received from one another (in contrast to other members of the wider farming community). Within the group, trust also enabled honest criticism of one another's business. This was essential for Grasshoppers' ability to fine-tune and adapt the New Zealand system, and at the same time in this case it helped to avoid the trap of groupthink where trust and community can lead to the uncritical reproduction of a shared way of seeing the world, a key asset in adaptive management (see Chapter 2). Instead, the values that are conserved through this supportive community were a tolerance for risk taking and innovation, and an openness to new ideas, even those that challenged individual perceptions and led to modified practices, the essence of organisational adaptive capacity. This was perhaps best shown in the expressed willingness of members to move from the New Zealand system to other solutions if the economic or environmental consequences of climate change required it.

The reciprocal of trust is exclusion suggesting the social limits of adaptation. In the case of Grasshoppers, exclusion was particularly strong around alignment with the culture of open criticism of farming practice. This could result in a personal challenge. The cost of membership is maintaining group standards, and dealing with group dynamics:

> I'd have to admit that at some points I've had to ask 'Is this worth the extra hassle? Do I need to be a member of this thing?'. But if you look at it in the longer term, I suppose everybody goes through points when they're extremely keen, and then not so keen.

In a network, where difference is positive because it enables exchange, exclusion is more likely to arise externally. In the workshop, there were several references to communication initiatives by the group in the UK that had not fared well.

It should not be assumed that trust is an unalloyed asset, and exclusion a constraint on adaptive capacity, or vice-versa. While it is certainly true that the learning culture within Grasshoppers had arisen through close ties of trust, it clearly also depended on exclusion. After all, potential members who could not cope with the group culture were expected to leave. Similarly, while exclusion enabled trust and learning, the question is what opportunities for learning and for wider social equality are being passed up in the name of maintaining group cohesion?

During the workshop, the group was optimistic about their ability to adapt to the challenges of climate change and variability, as and when needed. When pressed about this confidence, they ascribed it to successful change in the past: 'Having initiated change, it wouldn't bother us to change again in whatever direction, if it made sense.'

The adaptive capacity of Grasshoppers seems to be founded in a learning culture. The group fostered learning amongst its members, and this brought significant rewards for the effort of remaining an active member:

Discussion groups are the best way of learning – you can get to know each other's businesses, better than a lecture theatre.

It's like 20 heads learning at once, and sharing that information back. It would have taken me a lot longer to get where we are today.

The learning culture resulted in and was supported by a set of learning practices on the part of individual members, reinforced by the group's values. These had already built a culture of resilient adaptation to climate change:

We measure ground temperature and climate a lot more than other farmers. When we see change we change our practices. The data we have seen is getting warmer. The response to this is to withdraw fertilizer and put cattle out earlier.

A notable feature of the culture of Grasshoppers was the willingness of members to change embedded practices to achieve important life objectives, even to leave dairy farming. This seems a strong contrast with many other farmers who feel stuck, unable to make or even see the changes they need to remain viable. This also suggests that success in applying adaptation as resilience provides confidence for transitional and potentially transformational forms of adaptation at the level of individual businesses.

The members of the group were happy to view the Grasshoppers organisation as something transitory. The formal organisational structure was useful for the moment, but not necessary of itself. Seen as more important, and likely more enduring, were the informal relationships that group membership had fostered. This suggests that the shadow relationships that thicken the social ties within Grasshoppers now also prove a flexible social resource for forming new coalitions as future climate change and other challenges arise. While canonical organisation provides structure to help resolve defined adaptation challenges, shadow systems are the raw resource that should be strengthened to provide capacity to adapt to future uncertain threats and opportunities of climate change. That shadow systems are developed around and as a response to canonical organisation suggests a symbiotic relationship. This also points to a policy opportunity where shadow systems of relevance to wider society can be fostered through canonical organisations.

Conclusion

These two organisational case studies both show the idealised nature of resilience as a form of adaptation. Over time and faced with new challenges, policy directives or sources of information organisations are in a constant process of reinvention. Those that are not will likely not survive long in the everyday cut and thrust of market and political life. Given this reality there is a danger that rather than organisations being tools for protecting valued functions they strive to maintain their own longevity: resilience being transferred from function to form. Neither organisation studied here fell into this trap; members of Grasshoppers in particular observed that they valued the social bonds made through the group more than the group itself and that these were a resource should future challenges require new coalitions and communities of practice be formed. The close ties of trust in Grasshoppers also provided a key quality control mechanism that was less easy to observe functioning in the shadow system of the Environment Agency. In the latter case new ideas succeeded better with external validation – through academic papers, for example. The need for accountability and measured decision-making in public sector bodies is a particular challenge to those who would argue for embracing the shadow system to build adaptive capacity.

Respondents in both groups described their social relations in terms of communities and networks. Communities provide a powerful focus of social identity, but without the linking function provided by networks they risked becoming isolated from the broad pool of human learning. Networks, on the other hand, can be too diffuse, failing to provide an adequate basis for organised action, except in circumstances where the need to do so overrides the transaction costs involved in negotiating different interests.

The empirical observations made in this chapter support arguments from adaptive management for the contribution of relational qualities such as trust, learning and information exchange in processes of building adaptive capacity. They also caution that social networks or communities of practice will always exclude some actors and ideas and should not be seen as a panacea. For organisational management concerned with adapting to climate change four questions are raised by this research:

- Can the informal social relationships of the shadow system be embraced inside public sector organisations or are potential conflicts with the need for efficiency, transparency and vertical accountability intolerable?
- To what extent can investments in local formal organisations, like Grasshoppers, foster and maintain independent but linked shadow systems providing a secondary local social resource for climate change adaptation?
- To what extent might contingency planning to manage climate change risk compromise or complement efforts to build adaptive capacity to manage uncertainty?
- What management, training and communication tools exist to facilitate the building and maintaining of constructive social capital and social learning within and between organisations?

Modifying formal institutions to support motivated professionals in developing informal social ties and expand their membership of communities of practice to cross epistemic divides is one way of addressing this final challenge. At a larger scale investment in boundary organisations and individuals will help thicken the social resource for adaptation, and better cope not only with the direct impacts of climate change but the more dynamic organisation landscape that may well be an outcome of the economic as well as environmental instability associated with climate change. Scope for adapting governance regimes through transitional and transformational change is the focus of the next two chapters.

7 Adaptation as urban risk discourse and governance

In Cancun the most common idea is that 'it is not my problem, if things go bad, I can flee to another state'.

(Ex-member of the Quintana Roo State Congress)

The population mobility that enables and characterises rapid urbanisation has consequences also for discourses of responsibility, and finally the willingness and capacity of officials and those at risk to take action and reduce exposure and susceptibility to climate-change-associated hazards in a specific place. Mobile urban populations and the dynamic economies and social systems they are part of present both a context for climate change adaptation and, through the inequalities they generate, a target for transitional and transformational reform.

This chapter uses urban cases because the social and political concentration of urbanisation brings to the surface competing visions and practices of development. But the key argument of this chapter – that as discourses of adaptation begin to emerge worldwide they can either challenge or further entrench development inequalities and failures – is applicable across all development sectors.

Evidence for the interaction of adaptation with development norms and practice is presented from four rapidly expanding, but contrasting, urban centres in the Mexican state of Quintana Roo: Cancun (population in 2008 approximately 1.3 million), Playa del Carmen (100,000), Tulum (5,000) and Mahahual (1,000) (see Figure 7.1).

Quintana Roo is amongst the most rapidly urbanising places worldwide. Urbanisation is driven by state-sponsored and globally-financed international tourism in an area exposed to hurricanes and temperature extremes. National policy to exploit the environment of Quintana Roo for tourism attracts over 2 million tourists a year alongside large numbers of labour migrants from neighbouring states as well as international capital, and so generates risk to climate-change-associated hurricane hazards and more indirect impacts of climate change on the global economy and subsequent tourist numbers. As capital investment in tourism increases so the environmental attractor for tourists has changed from reef diving to beach tourism and now golf course condominiums. At each stage capital has inserted itself ever more forcefully between nature and its consumer. In so doing capital has generated and extracted greater financial returns while

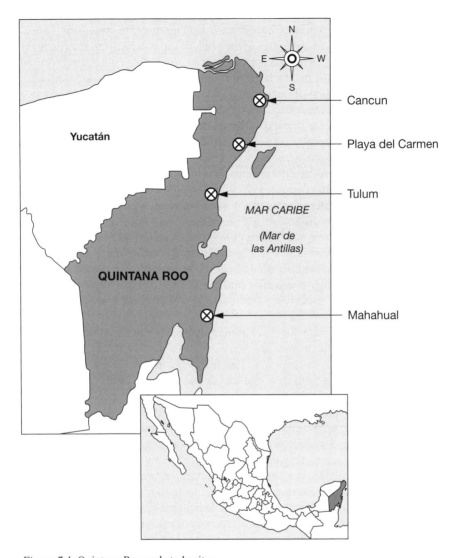

Figure 7.1 Quintana Roo and study sites
(Source: adapted from Cuéntame . . . de México, 2009)

alienating the consumer from her ecological foundations. The process has shifted economic reliance from a natural to an increasingly artificial 'second nature' (Smith, 1984). The result is a bifurcation in development strategies between those that exploit residual 'natural' spaces and the growing, capital intensive exploitation of second nature with greater environmental and social external costs as well as wealth generating potential. Capital insertion and the imposition of a second nature have occurred at different paces and can be found existing to varying degrees along the Caribbean coast. Cancun is the most intensive, with Playa del Carmen also presenting a mature capitalised urban system. Mahahual and Tulum are small urban centres at the brink of rapid capitalisation. The focus of this study is to explore the character of civil society within each urban form and so to examine the ways in which the urban process has given shape to and been influenced by this aspect of governance with a view to applying this knowledge to assess capacity to cope with current and adapt to future climate change impacts.

Layered on top of the impacts of capitalisation on the root causes of vulnerability and adaptive capacity is a more superficial but nonetheless important policy realm of hurricane risk management. This is the most tangible expression of hazard liable to be influenced by climate change on the coast. Records for hurricane activity in Quintana Roo begin in 1922 with a category one event (149 kilometres per hour). The first category four event was Charlie in 1951 (212 kilometres per hour), and it has been followed with increasing regularity by four additional category four hurricanes, and Gilbert (1988) and Dean (2007) both making landfall as category five hurricanes. These events reveal underlying vulnerabilities. Hurricane risk management succeeds well in compensating for proximate causes of vulnerability through evacuation of those at risk. But discourse around risk stops here, masking underlying root cause drivers of risk and unsustainable development.

This chapter contrasts with the empirical analysis of organisational adaptation presented in Chapter 4 both in terms of the scale of analysis but also the analytical lens. This shifts from one that stays close to the systems-based analysis of social learning and self-organisation to one that deploys aspects of discourse analysis and regime theory to help emphasise the political and value rich contexts that, alongside capacity for self-organisation, help determine innovation and dissemination of new ideas from the base and how far these might re-shape local governance regimes for adaptation and development in these sites. Different actors are shown to hold contrasting and sometimes conflicting visions of development that in turn lead to preferences for resilience, transition or transformation in society when faced with climate change. Following this introduction a short contextual section provides geographical and methodological background to the study. Each settlement is then analysed using a common framework with a concluding discussion drawing out contrasting relations between adaptation and development in each case.

Context: policy and methods

In 2007, the federal government launched a National Strategy on Climate Change and is now preparing a Special Programme on Climate Change to implement identified reforms. Thanks to these efforts Mexico has jumped from 14th in 2006 to 4th in 2008 out of 56 countries ranked according to their climate change performance in the Germanwatch, Climate Change Performance Index (Germanwatch, 2008). At the state level, while Quintana Roo's rapid demographic growth and infrastructure expansion open exciting opportunities to build climate-proofing into development, and at the same time provide a market edge around notions of climate friendly tourism, regional government and private developers have been slow to recognise climate change. In the language of transitions theory (see Chapter 4) this is an example of landscape (national/international) change meeting resistance at the regional (state) level. This begs the question: have any niche (local) level innovations emerged that might provide impetus for change at the regional level given the opportunity for change opened by perturbations at the landscape level?

Local impacts of climate change are felt already through perceived increases in the frequency of hurricanes, creeping sea-level rise, coastal erosion and high temperatures. These hazards are interrelated and compounded by local land use which has led to accelerated mangrove and interior deforestation, pollution and damage to in-shore reefs and the neglect of green and blue space in urban design. In contrast, state and federal agencies have a good record in containing human loss to hurricanes through timely if reactive strategies of early warning, evacuation and reconstruction of critical services. The most recent event, Hurricane Dean, 2007, caused limited economic impact across the region but made landfall close to Mahahual, which was severely damaged.

To reveal the values, capacities and actions of political actors in each urban centre an action research methodology was employed. In each settlement interviews were conducted with 12–15 leaders of social, environmental and business associations, and where formal organisation was absent amongst informal leaders. Following interviews, respondents were invited to town-level workshops to discuss results. Workshops provided an opportunity to verify reported views and interpretations, and also a vehicle for social actors to network. This was often the first time social actors had met to discuss climate change. A final workshop brought selected respondents from each settlement together to undertake a participatory comparison of town-level findings and again to provide a networking forum. Interviews and workshop texts were transcribed and data extracted and organised around the themes of development narrative, climate change, social-learning and self-organisation. Results have been fed back to civil society and government actors.

For additional material and analysis see: http://www.kcl.ac.uk/schools/sspp/geography/research/epd/projects/hslgmc

Case study analysis

For each case development pathways and perceived interactions with climate change are described. Innovations – from nuance to outright alternative – are identified and the capacity of civil society actors to promote such innovations assessed.

Cancun

In 1974, Cancun became the first integrally planned centre for mass tourism in Mexico and it continues to act as a centre for mass tourism, providing a significant foreign currency source to the federal government and generating significant employment opportunities. The population has grown exponentially from 88,200 in 1970 to 1,135,300 in 2005, creating a huge population potentially at risk from climate change associated hazards. Since 1970 Cancun has been directly affected by Hurricanes Gilbert (1988) and Wilma (2005). Vulnerability is aggravated by the exclusion of burgeoning workers' colonies developed on communally owned ejido lands which lie outside of the legal jurisdiction of local governments to provide basic services. Dominant development is controlled by the interests of national and international corporate capital, with many politicians having backgrounds as leading local entrepreneurs. Environmental legislation and urban plans are in place but frequent amendments and the slow pace of bureaucracy allow business interests great flexibility and resilience in the face of legislative, economic and environmental pressures – while increasing the time and trans-actions costs for environmental and social actors seeking to question development proposals. For example, the sensitive Nichupté Lagoon is under constant development pressure; despite an Ecological Zoning Program, hotels and squatter settlements have been allowed. The vision of Cancun as a centre for extractive capital built in a previously unoccupied zone disconnects social actors from a commitment to place and long-term planning. Reflecting on his experience, a Pez Maya Reserve fundraiser noted that 'those who participated the least were local entrepreneurs; more than 70 per cent of funds came from overseas'. For hotel workers Cancun society is described as fragmented with traditional values of neighbourliness and a strong family replaced by consumerism and undermined by drugs crime, alcoholism and extremes of inequality. Set against this, migrant workers maintain close links with source communities; for example, by sending children to school outside Cancun.

Respondents were clear that climate change impacts were exacerbated by past, and ongoing, development: deforestation was associated with increased temperatures and reduced humidity and the development of dunes for hotels contributed to beach erosion. Hurricanes were also associated with climate change and seen as integral to the development history of the city with specific hotel developments being cited as having taken advantage of Hurricanes Gilbert and Wilma to extend land claims into protected mangroves and privatise beach fronts. The comment below from an independent journalist shows the variable nature of disaster management and the long-term psychological and cultural impacts when the state does not fulfil its responsibilities for civil protection and security.

I remember that when Gilberto came Cancun was going to celebrate the Miss Universe contest – this focused the attention of the government in Cancun. With Wilma, however, the people of Cancun were left completely on their own. Until the army came to help two days later, the city lived in complete chaos. We stayed 15 days without electricity or water. The most incredible was the contrast between the rapid recovery of the hotel zone and the slow recovery of the rest of the population. Wilma brought a lot of despair and demoralized the population. We have not still recovered from that. People stopped to believe. Still today there is not the happiness that used to be. All this happened because we were left alone.

More positively, after Hurricane Wilma a Climate Change Association of Quintana Roo based in Cancun was founded. This small organisation has worked to promote recycling and lobbies against deforestation within existing development narratives. Elsewhere, actors seek to promote climate change in primary school syllabi. Business associations see climate change in economic terms with hotel associations needing to respond to tour operators threatening to lower tariffs or business volumes due to the poor state of the beaches. For the engineering and architecture community, climate change presents opportunities with recent projects including semi-permeable parking surfaces but with limited support from government. Overall civil society actors see their scope for action as limited compared to government which has power to revise urban planning guidelines, or simply enforce those that already exist, and this understanding of the distribution of power acts to suppress civil action and limit outright critique or confrontation of the dominant capital intensive model of development. This is despite civil society actors recognising that an emerging development paradigm that takes climate change into account can be an opportunity to enhance social development, with this being a particular concern in Cancun with its high inequality.

Civil leadership faces powerful opposition, as an environmental lawyer explained: 'It is dangerous to litigate against some powerful groups. We have to be very cautious'. For local actors the everyday experience of living with crime, including organised protection rackets, governs people's ability to voice complaints or instigate change. Civil society groups cited this and the culture of Cancun society, which is described as apathetic (caused by a crisis of credibility in the authorities), lacking in identity (with a diverse migrant population) and community spirit (with individuals working hard with little time for social work or volunteerism), as the main barriers to organising critical alternatives. Mass media is politicised and commercial. Individual civil society groups, lawyers or engineering companies might be competent and independent but they work alone, and often in competition, preventing the formation of a coherent social body and vision. One respondent described this as having a lack of institutional infrastructure to promote learning and new practices for adaptation and miti-gation, arguing that even if people were willing, without this infrastructure changing behaviour would be slow if not impossible.

While the institutional framework for strategic innovation and adaptation was lacking in Cancun informational resources were in place. Federal state agencies (the Ecological Gazette of SEMARNAT was twice mentioned) provided scientific data available for scrutiny and that had been used by civil actors in local litigation or lobbying. The Supreme Court judgement that all documents and studies related to a development should be in the public domain had also been used by local groups to challenge developments. The Universidad del Caribe, Cancun, is a local source of promotion for sustainable tourism region-wide. Collaboration with local government has been achieved by Amigos de Sian' Kaan in the preparation of a good practice guide for hotels that covers planning and use. In this way technical and management reform have been achieved by civil society groups to support vulnerability reduction, but social, economic, political and cultural systems remained outside discourse and unchallenged.

Playa del Carmen

Playa del Carmen has a successful and growing economy based on international tourism and in 1994 became the capital of the newly created Municipality of Solidaridad. Since then, Playa has been amongst the fastest growing urban centres of Latin America (above 20 per cent annually) and at times the fastest in the world (Campos Cámara, 2007). In 2005 its population exceeded 100,000 inhabitants. Playa has experienced direct hits from hurricanes. The worst challenge came in 2005, with Emily and Wilma a few months later. However, there were no fatalities and the town recovered very rapidly. In fact, the local tourist economy benefited from the relocation of tourists from Cancun, which had been hit even harder by Wilma.

In Playa, the dominant development narrative emphasised personal economic advancement and reflected the control over the local economy held by corporate private sector. Respondents reported on a disjuncture between people and place. Residents felt they were here to 'make money', not to settle. The result was a lack of popular commitment to local development and for holding private sector and government actors to account, as one social development activist reported: 'there is a lack of civic pride and identity with place – people do not care about the city or even their house and street'. Respondents described Playa as embodying an extreme version of the American Dream, celebrating individualism and materialism, and short-term gain over long-term development.

Some described climate change as a symptom of a larger problem of consciousness and the alienating effect of rapid urbanisation; as one respondent put it, 'We increasingly behave like machines – we need to go back to our community and our roots'. More broadly climate change could be a vehicle to hold the government to account if citizens became more engaged in governance. Adaptation (and mitigation) was seen as a leverage point for existing social and environmental agendas with progress reported in specific sectors; for example, the Sustainable Coastal Tourism Plan, believed to be the first in Mexico, includes guidance on beach and mangrove management. Huge scope for mitigation in the hotel sector was recognised with minimal current use of alternative energy, water recycling and waste management. The Small Hotels Association of Playa del Carmen and the Maya Riviera explained that the high proportion of family run hotels in Playa makes the sector responsive to calls for environmentally sustainable practices.

Strategy for future adaptation included reinitiating local food production as a local livelihood resource as well as a means of making some independence from global markets. External knowledge and expertise was accessed by NGOs through supporter networks and commercial links and had been instrumental in successful legal challenges to developments made on environmental grounds including X-Cacel, X-Cacelito and the Ultramar Doc. These were important symbolic successes, demonstrating that enforcing environmental controls need not jeopardise local economic growth.

Civil society groups tended to operate as top-down advocates or satellites to the government–corporate-business policy-making core. One social development leader observed that 'organisations are closed – they inform only staff and families, there is little public communication about plans or opportunities'. This reflected the lack of trust and individualised nature of Playa's society, one where, as one respondent put it, there was 'no culture for donations, public participation or volunteerism'. Perhaps because of a lack of local embeddedness, the personalised character of civil society organisation and its orientation towards government, there were few examples of collaboration across sectors. This is a particular challenge for building capacity for progressive adaptation.

Individual acts had successfully challenged dominant cultural and social norms; for example, through the provision of civic amenities including the Ceiba Park to show local residents that they too, and not only tourists or the locally wealthy, were worthy of a healthy local environment. Speaking up in public consultations was claimed to have symbolic as well as instrumental significance through demonstrating the exercising of a local voice. The facilitating of neigh-bourhood talking groups aimed to strengthen families. Still many residents did not see themselves as citizens of Playa but of their home towns and states, making the building of any grassroots-led call for change very challenging. Greater capacity for adaptation, albeit of a resilient kind, was observed with civil society groups operating close to the private sector: innovation included dive companies that opened inland cave-diving sites in response to deteriorating coastal environments. The importance of local and global ecosystem services to the local tourist economy also provided a narrative for current development planning and regulation and one that could be adapted to include climate-proofing.

Tulum

Tulum was until the 1970s a Mayan ejido of subsistence farmers. The ejido's settlement was established about 2km inland, and thus protected from hurricanes by a generous stretch of mangroves and forest. Driven by in-migration, Tulum's population grew exponentially following the construction of a highway and in the 1980s, as low density hotels proliferated along its outlying beach front. By 2004 there were 53 hotels in Tulum offering 1,235 rooms and a permanent population of around 1,000. Hotel designs range from concrete three-storey build-ings to very basic thatched cabins, and often include renewable energy and other eco-friendly features. Although some of the hotels are owned by external actors (Mexican and foreign entrepreneurs), the majority are owned and managed by local or partly local entrepreneurs. As in Playa, ejidatarios benefited from selling land and some of them are now wealthy even if still preserving some of their traditional ways of life. In April 2008 the state government granted the independence of Tulum as a new municipality.

Today, Tulum is at a crossroads with two competing development narratives. The dominant narrative portrays Tulum as an opportunity for speculative development. This is symbolised by the 'Downtown Tulum' development, a project forged and implemented by Yucatecan entrepreneurs in concomitance with the governor of the state. The works for the first phase started in January 2008 and contemplate the urbanisation of 77 hectares located between the town and the beach. The second phase comprises 450 hectares including a mega golf course that would extend up to the beach and a grid of water channels resembling an inland marina.

An alternative narrative is oriented more towards local Mayan values and ecological and community sustainability. This vision was championed by a small group of well-educated local businesses leaders and civil society groups, but optimism for the future of Tulum as a sustainable tourism centre is limited. Respondents presented striking visions for an alternative development, but felt in reality small gains that can build resiliency into development are all that is likely to be achieved in resisting the corporate transformation of Tulum.

> We already have failed models such as Acapulco and Cancun and we do not want to fall into the same in Tulum. It is almost impossible for local people to affect the model or direction of development. However, there are local pressures to, for instance, make wider sidewalks or guarantee the connection of drainage to a waste water treatment. We want that they build drainage before paving any street. (Former president, Tulum Hotel Association)

Alternative economic vision is provided by community (ejido) owned development at the Dos Ojos cave system and at a bio-region project at Jacinto Pat Ejido. Most ejido lands and individual owners have sold to speculative capital and subsequently left Tulum but these examples show an economic rationale for development led by and for the benefit of local people with a concern for environmental integrity. Some medium-scale migrant entrepreneurs support this vision, with the Chan Chay Ecological Shop providing green cleaning products for the hotel sector but also organsing workshops, and a Green Expo in Tulum to promote this site as a 'green spot on the Maya Riviera'.

Local consequences of climate change are recognised, most significantly associated with increased hurricane activity and higher temperatures, both exacerbated locally by deforestation of mangroves and coastal forest and intensive urbanisation.

> Climate change is impacting through housing development. The areas that were for conservation are now being urbanized and this is generating

disequilibrium. There have been protests against the Aldea Zama project and now we have an environmental department in the Municipal Council, but this type of progress is screwing us up. (Manager, Zero Workshop Foundation)

Tourist occupation as well as local quality of life is reported to be affected directly, with concerns on a shift from high- to low-end tourists at periods of hurricane activity, and indirectly; for example, through the loss of a section of coast road with every passing hurricane. For other civil actors climate change presented an opportunity to press dominant development processes and lobby for change. For the Centro Ecológico Akumal 'climate change can slow down development due to the recurrence of hurricanes. This would give us a chance to shift the dynamics'. Similarly, the Chan Chay Ecological Shop saw climate change and its media coverage as contributing to ongoing efforts to motivate individuals and businesses to become more ecologically responsible. For most civil society actors capacity to respond was limited to raising awareness through public workshops and school visits. Coastal reef management has generated some local research and conservation work receives international coverage but is not framed by climate change adaptation.

Social leaders identified considerable barriers to organising for adaptation and change. There was no culture of active resistance in Tulum, but rather one of silence and compliance; at the same time new migrants were less concerned about Tulum's environment than economic opportunities and so supported the dominant vision of development. For the local and migrant populations compliance was underwritten by a lack of educational opportunities, with TV being the primary source of information and opinion forming. Several respondents saw the promotion of an alternative development not as a challenge of providing information but of working with partners to raise critical consciousness – a deep shift in local mindsets that are accustomed to mediating development through adaptive ingenuity, to use Freire's terminology; an ambitious aim and one made more so by the weaknesses of the civil sector in Tulum, which was acknowledged to be small with isolated organisations easily coopted by dominant business interests. Middle classes and young professionals that might be at the forefront of organising local social movements were overworked and had little time for public work. The crisis in leadership was such that some respondents looked hopefully to international NGOs.

While undermining local visions for development, the urbanisation process itself also offered opportunities for organising alternatives. Development increased the external visibility of Tulum, and provided opportunities for accessing information; for example, through technical support from the federal agency SEDESOL and the French Embassy, which offered knowledge exchanges with French Municipalities. Drafting of the Urban Development Plan included citizen consultation but with limited impact, with the most positive consequence of this experiment in participatory governance being its slowing down development – providing time for alternative discourses to assert themselves outside of the formal planning process. The new municipal authority expressed concern about the loss of Tulum's existing cultural and ecological character in the tidal wave of approaching development, but looked to the federal government for leadership and capacity.

Mahahual

Mahahual is a pioneer settlement with a population of about 500, largely in-migrants from Mexico and internationally. From 2008 Mahahual was conferred the status of Alcaldía and administered through a local council with responsibility for the tourism centre with its beach properties, modern residential properties, cruise ship terminal and several small satellite residential and farming communities, including an informal settlement located two kilometres away from the main centre. As the economic base shifted from fishing to tourism rapid in-migration and land speculation have changed the physical and social structure of the town. Few original families remain and these are a small minority compared to the immigrant population. The local economy has experienced a boom since the construction of the cruise ship terminal, with land speculation driving a healthy virtual economy. Hurricane Dean made a direct hit on Mahuhual in August 2007, with the subsequent closure of the cruise ship terminal stalling the local economy.

As a pioneer settlement there was a feeling of excitement and opportunity directed by a desire to build Mahahual without being dominated by cruise tourism. It was an 'open frontier' where local residents had a central stake in shaping the future. Here, the need to build community was a common aspiration with some working towards this, but mistrust in social organisation and leadership was pervasive, in business, social and local government organisation alike. Environmental concerns were marginal; for residents development meant the improvement of critical physical and social infrastructure and promotion of the local economy. However, one leader of a social development group suggested that following Hurricane Dean a slow process of cultural change may have begun: 'after Dean one is starting to feel more solidarity. It is happening as in Cozumel, people there are building solidarity as a result in part of facing many hurricanes'.

The common construction of climate change in terms of hurricane risk played down long-term thinking. Accepting hurricane risk as a development externality also contributed to individual businesses and the regional and federal state being cast as the actors with primary responsibility for responding to climate change. The local council, which should be a driving force for adaptation, had not yet taken this role. Practical action was limited to associated environmental agendas; for example, the Tourism Entrepreneurs Association of Costa Maya campaigned to clean the village with the participation of the authorities, and lobbied to prevent trucks coming into the village and for investment in waste recycling. Information networks were extensive stretching to other parts of the state, Mexico and overseas, and led, for example, to calls for a local civil protection body in local government.

Before Hurricane Dean, low levels of trust with any form of social organisation was aggravated by Mahajual's diverse and atomised society, with many
immigrants and a small population base that constrained the leap from individual
to collective action. The leader of a fishing cooperative reflected on the impact of
low trust on the formation of his group: 'We had to make three meetings before
we could elect a president. People tend to attack those who stand out from the
rest. They think one is looking for his own benefit.' The combination of economic
and governance constraints was exemplified well by the residents of Km55, a
satellite settlement with formal and informal land holdings where one leader
reported that 'only 36 of 400 plots are occupied, the rest are held speculatively;
this makes it hard to organise'. Another noted that 'uncertainty about land titling
is delaying; for example, people will not put electricity in their lots until this is
solved'.

After Hurricane Dean, reconstruction opened a window for building common
identity (as temporary labour migrants and uncommitted investors left) and
potential for collective action. A businessman reported that:

> Before Dean I tried many times to create an association, but without Dean
> and all this easy money nobody paid much attention. All the ideas that I was
> proposing turned out to be right after Dean. Now people are starting to build
> common culture because the ones who have stayed do not see this place only
> in terms of money.

Some individuals also took advantage of governance failures post-Dean with
examples of mangroves being illegally cleared, but for those seeing potential in
collective action reconstruction served as a common context for organising. A
sense that local civil society actors had a stake in shaping the future of Mahahual
was reinforced by a search for alternative tourist markets following the temporary
closure of the cruise terminal. This was driven by individual companies with
minimal state support. Still many respondents felt that Mahahual's recent
Alcaldía status would also open new opportunities for collaboration with local
government and the Alcadía was also concerned to project itself as seeking to
build partnerships with local civil society, providing real scope for mainstreaming
climate change.

Conclusion

The preceding analysis presented dominant and alternative discourses on development, climate change and scope for adaptation in each study site from the
viewpoint of local civil society actors. Here a comparative analysis is presented
to draw out differences in the ways in which adaptation was used to promote
resilience, transition or transformation within the particular development contexts
of each site. Table 7.1 summarises this analysis. As a caveat, it is important
to note the methodological challenges in capturing and then representing
the diversity of views on development and climate change in a reductive but

Table 7.1 Adaptation as an opportunity and narrative for development discourse and action

		Cancun	Playa	Tulum	Mahahual
Dominant development vision		Intensive, large scale, corporate extractive capitalism	Corporate and local extractive capitalism	Transform environment into commodity for speculative investors	Small scale pioneer capitalism
Perceived climate change risk		Translated into a challenge for tourism marketing, insurance and engineering design; not a concern for local social and environmental integrity	Risk of marginal concern in the planning horizon of businesses and government	Risk denied or assumed to be planned out in the future so of little consequence for future investments	Climate change threatens economic base through damage to cruise tourism
Adaptation opens scope for:	Resilience as discourse	Improve coastal engineering and tourist building design	Maintain beach and coastal water quality	An opportunity for greening business and promoting mitigation	Generate new markets independent of cruise tourism
	Resilience as action	Beach replenishment, artificial reef design, hotel retrofit	Beach replenishment, dive companies market interior sites	Marketing and informing businesses	Individual acts of marketing
	Lead actors	Municipality, engineering consultants	Municipality, SMEs	SMEs	SMEs
	Transition as discourse	Assert rights to police dominant vision by exercising entitlements for environmental sustainability	Assert rights to challenge dominant vision by exercising entitlements for development control	Economic growth is welcome if controlled	Assertion of identity through new council status and following Dean to leverage funds for local development
	Transition as action	Engage in development consultation and take legal action	Legal challenges prevent developments	Engage in citizens consultation for Urban Development Plan	Collective acts of reconstruction after Hurricane Dean
	Lead actors	Environmental NGOs and lawyers	Local environmental NGOs and Cancun based lawyers	Some local civil society organisations	Local council, SMEs

Table 7.1 Continued

	Cancun	Playa	Tulum	Mahahual
Adaptation opens scope for: Transformation as discourse	Call for extension of basic needs and risk management to migrant worker colonies; puts distributional equity at the heart of alternative vision	Building self-worth and critical consciousness amongst migrant workers as a first step for reclaiming a voice in development	Raise critical consciousness of environmental and cultural costs of extractive development	None
Transformation as action	None	Symbolic acts, e.g. La Ceiba Park reclaims quality green space for locals	Popular education	None
Lead actors	Independent journalists and social development NGOs	Social and cultural development NGOs	Cultural NGOs	None

meaningful way are not insubstantial. Table 7.1 seeks only to represent the most influential narratives and associated actions and key actors linked to resilience, transitional and transformative adaptations. Resilience is indicated by efforts to maintain business-as-usual development paths; transition exercises existing legal and governance rights to confront unsustainable development, and transformation uses adaptation to promote fundamentally alternative forms of development from those described for each site as dominant.

Across the sites some commonalities emerge. Local government and business interests are prominent in responding to climate change through building resilience, which is also the predominant form that adaptation takes in each case. In contrast civil society groups and environmental lawyers are most prominent in transitional acts, using adaptation to push for greater transparency, participation and accountability within the existing governance system. Cultural actors, including NGOs and journalists, emerge as leading transformation, which exists largely at the level of discourse, with some acts of popular education and symbolic initiatives aimed at promoting popular critical consciousness. Given the strong voice of government and business in shaping the limits of adaptation it is perhaps not surprising that ecological modernisation is the dominant overarching worldview within which adaptation is being constructed as resilience (from coastal engineering in Cancun and Playa to the greening of business in Tulum), and transition (the use of legislation to regulate development in Cancun and Playa).

For individual workers coping with risks, including those associated with climate change but driven more by a search for economic opportunity, is played out within the use of migration as a livelihood strategy. Emotional commitment to locales in Quintana Roo is spread thin and legitimised through cultural norms that accept local residence as temporary and extractive. In contrast migrant workers maintain close links with places of origin, even sending children 'home' to be educated. This offers an opportunity for individual and familial resilience with low social transactions costs – without the need to engage in social or political collective action in the place of residence.

Given the general acceptance that climate change is already impacting negatively through beach erosion, high temperatures and hurricane activity the level of proactive planning is minimal. This may be a function of the linking of climate change with environmental management and subsequent policy marginalisation, but possibly also points to a denial of risk, especially by those most vulnerable. The common tendency amongst the poor and vulnerable to prioritise economic opportunity over risk reduction is heightened through a majority migrant population that has little association with place or community. Corporate interests in Cancun and Playa have access to engineering solutions and international insurance, and beyond this possibly view their investment in Quintana Roo as temporary. For smaller businesses and the resident population scope to adapt is more limited, and as was most keenly demonstrated in Tulum, for many migrants rapid transformation of the environment into a form that can be exploited by capital has attracted them to the coast. Climate change is pushed to the margins of people's imagination as well as their actions. The one major exception is

Mahahual where Hurricane Dean caused the loss of the town's economic base. While Mahahual's population is almost entirely composed of recent migrants, the effect of Dean as well as the recent awarding of town council status has begun to build a social identity.

The aim of this chapter has been to reveal the messiness of analysing adaptation where political values and actions are both contested and tightly circumscribed by a rigid political and economic framework. In the language of transitions theory the cases all display strong tendencies for stability with limited scope for local innovations to affect change in regimes through adaptation, partly a result of the limited range of innovations observed (examples included the La Ceiba Park in Playa, which provided the dual function of meeting a service need for urban green space but also potentially inspiring critical consciousness, and material alternatives such as ejido controlled development and the Chan Chay Ecological Shop in Tulum). This is compounded by a lack of a supporting institutional architecture (including values and a legal–administrative framework) to aid the dissemination of innovations; and a strong dominant existing political-economic and administrative regime. Even where disaster events have been experienced, revealing failures in the dominant regimes and development pathways, pre-disaster political, economic and cultural structures have changed little. Resilience remains the dominant mode of adaptation across this region. It remains to be seen how far this will be true as increasing population, physical and financial assets are exposed to climate change associated hazards in the future.

8 Adaptation as national political response to disaster

> . . . moments when underlying causes can come together in a brief window, a window ideally suited for mobilizing broader violence. But such events can also have extremely positive outcomes if the tensions [. . .] are recognized and handled well.
>
> (USAID, 2002)

This description of post-disaster political space highlights the possibility that political outcomes are not predetermined by history but open to influence, in this case by the interests of an international political and economic actor.

Context: policy and methods

The reflexivity of socio-ecological systems allows us to envision climate change impacts as unfolding within ongoing socio-political trajectories. Disaster events, and especially reconstruction periods, open space for change in dominant technical, policy and political regimes (Pelling and Dill, 2010). Very often such changes are best classified as adding resilience to pre-disaster socio-technological systems. New technology to improve the resistance of infrastructure, or policy reform such as the enforcement of building regulations, allow political and economic business as usual. Sometimes, however, unacceptable failures in the dominant regime to meet its responsibilities for risk reduction and response can act as a catalyst for political level change and open scope for transformational adaptation that goes beyond disaster risk management to influence social life and the distribution of political power in society. Chapter 7 identifies the most likely pre-conditions for such changes, which include economic inequality, a pre-existing and organised alternative to dominant politics and a sufficiently high impact event (Albala-Bertrand, 1993; Drury and Olson, 1998; Pelling and Dill, 2006).

It is not only natural disasters that provide sufficient shocks to destabilise dominant political regimes, but these are perhaps the most directly related to the influence of climate change. In the future, climate change will likely be a factor of growing significance in many other kinds of shock, especially those compound events felt locally from the conjuncture of multiple factors. The most recent example of this was the 2008 global food crisis. A combination of changes in local planting regimes (a shift from wheat and maize for consumption to bio-fuels), increasing demand (for example, from China's rapidly expanding middle class), exceptional drought and the failure of key regional harvests (for example, the Australian rice harvest), and instability in the global financial systems (commodity speculation at a time of high carbon fuel price) destabilised water-food systems resulting in increased hunger and malnutrition for the poorest with crises in 37 countries. At places this has fed back into the political system through violent protests in such diverse countries as Cameroon, Egypt, Haiti, Indonesia, Mexico, Morocco, Pakistan, Senegal and Yemen (FAO, 2008). In this context, natural disaster events provide early insight into the ways in which specific political systems respond to shocks and what we might reasonably expect if failure to adapt to reduce risk leads to more frequent and severe events (Schipper and Pelling, 2006).

Case study analysis

This chapter presents three case studies. Each is summarised in Table 8.1. The first case study from Bangladesh unfolds in a period of post-colonial nation building; the remaining studies from Nicaragua and the USA occur in the contemporary period of globalising capital where political dominance is not simply concentrated in the state but more diffusely spread amongst national and international private sector and civil society interests. Each case is built around direct quotations from eyewitnesses or observers with comment on the pre- and post-disaster polity. The cases serve to illustrate that adaptation is more than a narrow technical activity, and can encapsulate the political as well. In doing so adaptation becomes a contested space that competing social actors attempt to capture at the level of symbol and discourse as well as through material actions. The final impacts of disaster events are difficult to describe as with passing time new events place their influence on political trajectories. Two possibilities have been hypothesised: a critical juncture (Olson and Gawronski, 2003) describes those moments that when passed cannot be reversed; in contrast an accelerated status quo (Klein, 2007) is felt when pre-disaster social and political relations are further entrenched through disaster. The core distinction between these models is between change as an outcome of the successful concentration (accelerated status quo) and contestation (critical juncture) of established political and associated economic and cultural power (Pelling and Dill, 2010).

1970, East Pakistan (Bangladesh): the Bhola Cyclone and the politics of succession

The Bhola Cyclone devastated East Pakistan (now Bangladesh). The failure of leadership from West Pakistan (now Pakistan) enabled the disaster to feed into an already popular succession movement and is a prime example of a critical juncture event.

Following two hundred years of British rule, East Pakistan was formed in 1947, governed by Western Pakistan, some 1,000 miles away. Despite their shared Muslim religious heritage, the populations of Pakistan's two territories had significant cultural differences with the predominantly Bengali population of East Pakistan enjoying close cultural relations with Indian Bengalis living near their border (Washington Post, 1971). Differences between East and West Pakistan became politicised during the nation building process; for example, through West Pakistani leaders insisting that Urdu (the lingua franca of West Pakistan) be instituted as the state language (Oldenburg, 1985). Against this background, a popular movement for cultural autonomy had existed in East Pakistan since 1947 and was given a political dimension by the political and economic disadvantages experienced by the Eastern province.

> In 1970, Bengalis were living in what would soon become one of the world's most densely populated nations. Land scarcity forced Bengalis to build homes in areas subject to recurring floods. Increasing numbers pushed southward to clear and settle the Sunderban Forest (what used to be the home to the Bengal tiger), and deep into the south coast, which exposed them to the vagaries of the Bay of Bengal. (Sommer and Mosley, 1973:120)

In 1970, a massive typhoon hit:

> On 12 and 13 Nov 1970, a cyclone and tidal waves hit Eastern Pakistan (now Bangladesh) resulting in colossal damages to both human lives and properties. Some 10,000 square miles, covering a number of off shore islands in the Bay of Bengal were affected. Total population affected was approximately 6.4 million and estimated death toll was in the region of 2 million. (MINDEF, 1970)

Soon after the catastrophe, a medical team from Dacca (Dhaka) interviewed survivors who described either a gradual increasing of flood waters over a period of hours or conversely, a sudden 'thunderous roar followed by a wall of water'. The team reported:

> Where the water rose gradually, people scrambled on to roofs of their houses or scaled trees. But the houses frequently gave way, and only the strongest could maintain their grip on the wet and slippery tree trunks in the face of the 90 mile-per-hour winds. In areas where the tidal bore struck suddenly, there was even less hope of withstanding the force of the waves. (Sommer and Mosley, 1973:122)

One witness to the devastation described the scene incredulously:

> Flying out to the Bay of Bengal 2–3 days later on persistent reports of massive casualties, the rivers flowing into the ocean seemed clogged by the

carcasses of animals and debris. Nobody believed us when we said these were corpses of human beings, in the thousands and thousands. The Islands of Hatiya and Sandip lost part of their population. Bhola and Manpura (and tens of smaller Islands and coastal areas like Kuakata) were swept almost clean of humans, animals and houses. (Sehgal, 2005)

The central government in West Pakistan was either unable or unwilling to act. Commentator Amir Ayaz suggests that both physical and social distance stayed the hand of the central government:

While a tidal wave of death and destruction swept over the eastern wing, the military government was slow to respond, paralysed by what I can only think of as a sense of remoteness. East Pakistan and its coastal people were just too far away. Which is a bit like the Bheels of Thar and the Koochis and other nomads of Balochistan. Mainstream Pakistan passes them by. Imagine if the water supply of Islamabad were to be closed for two or three days running. The howls of anguish rising as a result would touch the heavens. (Amir, no date)

When the government did finally act, its measures were limited to helping the least affected population, leaving the worst hit areas virtually abandoned. The medical team from Dacca (Dhaka) reported that:

While the minimal amounts of bamboo distributed by the government were adequate for repairing the roof or sides of a house in the more northerly areas, they were wholly inadequate for rebuilding the entire structure, which was necessary in the more devastated coastal regions. The results were pathetic: tiny grass and straw huts, three of four feet wide and high and perhaps six feet long, each housing a family of two to eight persons. (Sommer & Mosley, 1973:125)

Consequently, villagers in less affected areas were soon busy reconstituting the fabric of society, but this was not the case in the worst affected coastal regions. The team observed that:

There the men were usually found squatting despondently in the centre of the village. They lacked all the implements basic to achieving self-sufficiency, and they had no money with which to buy them. (Sommer & Mosley, 1973:127–28)

The Pakistani government's failure to adequately respond to the devastation of the typhoon gave East Pakistan's majority party, the Bengali Awami League, a stronger position from which to negotiate. The UNDP supported Sustainable Development Networking (SDN) project explains:

> [T]he regime was widely seen as having botched (or ignored) its relief
> duties. The disaster gave further impetus to the Awami League, led by Sheikh
> Mujibur Rahman. The League demanded regional autonomy for East Pakistan,
> and an end to military rule. In national elections held in December, the League
> won an overwhelming victory across Bengali territory. (SDN, no date)

In December 1970, just one month after the disaster, national elections were
held. The Awami League took all but two National Assembly seats reserved for
the eastern region, and was suddenly launched as a majority political force on
a par with West Pakistan's People's Party (Sen, 1973). Ikram Sehgal (2005)
argues that the popular moral outrage over the government's poor response to the
disaster catalysed the independence movement:

> The Federal Government remained distant, seemingly cold and unfeeling in
> Islamabad. The perception of little or no relief set the stage for far reaching
> adverse consequences. The cyclone brought the anti-Pakistan antagonism
> building up over the years to a head in such circumstances it was sheer mad-
> ness to go through with the scheduled November 30 elections. The political
> result was a foregone conclusion, a massive protest against the Federation, as
> it existed then, later became a mandate against the very continuity of Pakistan
> as a nation. (Sehgal, 2005)

The election demonstrated Bengali resistance to the continuation of martial law
and support for democracy and regional autonomy had coalesced into a powerful
political movement. But secession was apparently an act of desperation. Philip
Oldenburg (1985) asserts that the Awami leadership would not have been averse
to taking a leadership role in a consolidated Pakistan. He points to the fact that the
Awami League did not announce secession until the central government reacted
to the election results with massive violence. Robert LaPorte (1972) writes that
the West Pakistani reaction to the election was to conduct 'ethnic cleansing'. He
explains that in order to crush the autonomy movement, the central government
acted to rid the so-called 'misguided' Bengalis of the forces that were breaking
up the nation. Thus the state proceeded to arrest or kill Awami League leaders
leading to a massive exodus of Bengalis to India, and ultimately to India's
decision to engage its army to back the Bengali war of succession. In April 1971
the exiled government took oath with Tajuddin Ahmad as the first prime minister.
Sadly independence did not free Bangladeshis from exploitative government,
political violence or natural disaster. Nationwide famine struck in 1973 and 1974
(Sen, 1981). Coups, assassinations and claims for one party states have distorted
national politics. Bangladesh is now considered to be one of the countries most at
risk to the impacts of climate change and her population is highly vulnerable to
riverine and coastal flooding as well as drought and food security.

Interestingly, the authors cited here whose work was published in academic
journals make no mention of the catastrophic typhoon in their analyses of the
events surrounding Bangladesh independence. Whereas the authors published in
public forums (NGO report and OP Ed, respectively) write as though the connec-

tion between the failure of the central state to provide for the population following the typhoon, and increased resistance to West Pakistan rule was patently self-evident. It is likely that the relative newness of treating environmental crises as politically significant events, combined with an academic avoidance of anything that could be perceived or misunderstood as environmental determinism, explains why the disaster did not figure in the analyses of the former.

In summary the disaster, set into motion by socio-political policies that forced Bengalis to live in conditions of high vulnerability, swelled the ranks of the discontented and radicalised many. Pakistani state violence against Bengalis, linked to the dominant ideology of the homogenous nation-state, effectively closed space for Bengali manoeuvrability. The disaster pushed popular sentiment towards support for a war of secession.

1998, Nicaragua: Hurricane Mitch, a missed opportunity for transformation

Hurricane Mitch exemplifies resistance in the social contract before and after a catastrophic event. Political interests both generated vulnerability and risk in Nicaragua and diluted the promise of the reconstruction period which was presented as an opportunity to break from the past and turn reconstruction into a transformative development moment. Despite transformational rhetoric including the decentralisation of development governance after Mitch, material, progressive change has been limited: a missed opportunity for adaptation to enhance progressive development.

The contemporary history of Nicaragua is eventful and dramatic. Michael Pisani neatly summarises some of the extraordinary socio-political shocks sustained by Nicaragua over the course of just 25 years:

> It is difficult to discuss present-day Nicaragua without describing the astounding transformation that has taken place in the country over the past generation. In brief, these extraordinary events and changes include 1) insurrection and popular revolution, 2) counter-revolution and low-intensity warfare (the Contra War), 3) 100,000 dead as a direct or indirect result of armed conflict (2.5 percent of the population) and a halving of national output, 4) a period of hyperinflation that reached an annualized 33,000 percent in 1988, 5) socialisation of the economy, 6) privatisation of the economy, 7) debt crisis including a 1990 per capita foreign debt figure of $2,867 in which per capita GDP was $469 or a foreign debt-to-income ration of 6.1 to 1, 8) seven national leaders (1979–2002), and 9) three debilitating natural disasters (the omnipresent 1972 earthquake in Managua and two destructive hurricanes, Hurricane Mitch in 1998 and Hurricane Joan in 1988.
> (Pisani, 2003: 112)

Given the role that Anastasio Somoza Debayle's mishandling of the 1972 earthquake reconstruction played in preparing the Nicaraguan population for popular insurrection (see Table 8.1), it is not surprising that the Sandinistas moved quickly to improve the national system for disaster mitigation and response after gaining power. The Nicaraguan Institute for Territorial Studies (INETER) – which currently houses state scientists in geology, meteorology, geophysics departments, produces the nations' maps, registers land and provides data for land use policy – was created by legislation signed by the Sandinista government in 1981. In 1982 the government transformed the Nicaraguan Civil Defence into a nationwide network of civilians dedicated to promulgating the revolution amongst the Nicaraguan populace while the US funded Contras attempted to topple the government through low intensity warfare: a clear case of adaptation combining a technical and political ambition. However, the Civil Defence was not merely a propaganda machine; by the late 1980s Nicaragua for the first time had a cadre of at least rudimentarily trained emergency managers (Olson *et al.*, 2001).

In 1990 a peaceful transition in power saw a landslide victory for the neo-liberal National Opposition Union. The new government allowed the Civil Defence to continue its functions proving effective during the Pacific Coast tsunami of 1992, and, after European Union funding helped link the organisation with the scientists of Nicaragua's Institute for Territorial Studies (INETER, 1998), it performed especially well during Hurricane Cesar (Ibid.). Thus, in Nicaragua, three of the basic elements fundamental to effective disaster mitigation were ostensibly already in place when Hurricane Mitch battered the isthmus: 1) a national institution housing earth scientists and providing early warning; 2) an established national network of civil defence; and 3) an organised citizenry accustomed to working with civil defence. What happened?

> What turned Mitch from a natural hazard into a human disaster was a chain reaction of social vulnerabilities created by long-term climate change, environmental degradation, poverty, social inequality, population pressure, rapid urbanization and international debt. (Rodgers, 1999)

> Multinational companies financed many of the coffee plantations neatly terraced into the mountainsides of Nicaragua and the banana plantations cared out of the lush coastal regions of Honduras. Both types of plantations were viewed as beneficial economic enterprises but they had the secondary effect of displacing small farmers further into the mountains where they in turn cut down forests to grow subsistence crops[. . .]But the long term environmental consequences of clear-cutting land for agricultural purposes were never anticipated in the region's development plans. Potential economic losses were never calculated, nor were mitigating actions taken to reduce the harmful effect of erosion. (Comfort *et al.*, 1999:40)

Comfort *et al.* (1999) identify austerity measures required by externally imposed structural adjustment programmes as the impetus for cutbacks in public services such as health and transportation, which are in turn responsible for reductions in the capacity of local and national governments to respond effectively to the disaster. Olson *et al.* (2001) provide evidence of how these policies affected Civil

Defence where almost half of the 58 officer positions distributed across seven regional offices were not filled.

From October 21–31 1998 the western and northern coasts as well as the central region of Nicaragua experienced from three to five times the rainfall ever recorded. According to the Nicaraguan government Hurricane Mitch destroyed or damaged 151,215 homes, 512 schools, 140 health centres, 5,695 roads and 1,933 bridges. The government confirmed 3,045 people dead as a result of the disaster (Olson *et al.*, 2001). These human and material losses occurred in a country with a population of only 4.5 million people. Two thirds of the total fatalities associated with the storm occurred in one ghastly 'disaster within a disaster':

> The single most horrific event occurred in Nicaragua on October 30, 1998, when the side of the Casita volcano collapsed. Loose volcanic ash accumulated from centuries of eruptions became a deadly flow of mud and debris known as a lahar. During the night, it hurtled downhill at speeds of up to 60 miles an hour for seven miles, burying 2,000 people in the villages of El Porvenir and Rolando Rodriguez. (USAID, 2005: 4)

The magnitude of the national disaster was not due to a lack of an early warning system. Meteorologists from INETER tracked and duly reported the location and acceleration of the storm (INETER, 1998), providing the government with the information it possessed in a timely and efficient manner. But this was not sufficient: knowledge of the hazardous geology and population at risk was not available and had certainly not been acted on to reduce development of this area (USGS, 1999). Risk was produced as a result of a combined lack of appropriate scientific knowledge and an underlying political economy that allowed, or forced, the poor to colonise a hazardous location. Political inaction aggravated the impact of Mitch. Despite the enormity of the environmental phenomenon taking place, and several days into what was becoming a regional catastrophe, President Alemán failed to act on government ministers' advice that a state of emergency should be declared and evacuations and rescue missions organised. In an essay published shortly after the disaster, the Director of the Nicaraguan Centre of International Studies, Alejandro Bendaña, suggests that one reason President Alemán refused to initiate a massive, organised emergency operation to mitigate the effects of the storm was that this sort of action would be reminiscent of the populist campaigns conducted by his political nemesis, the Sandinistas: 'No he said, such a mobilization would be something that the Sandinistas would do – and he certainly was no Sandinista' (Bendaña, 1999).

The president's refusal to respond appropriately to the needs of the Nicaraguan population before and after the storm was judged by international analysts as likely to result in political fallout:

> Assessments of the political impact of the hurricane are necessarily highly tentative at this stage. However, early indications suggest that in the medium term, the disaster may lead to an increase in popular opposition to the

government of President Arnoldo Alemán Lacayo. For several weeks before the hurricane struck, producers had been calling for government assistance to help them cope with the impact of higher than usual rainfall through October. However, the government did not call a state of emergency until early November, after the hurricane had struck. This is likely to reinforce a growing sense among the populace that the current administration is indifferent to popular sentiment. (EIU, 1998: 7, cited in Olson and Gawronski, 2003)

The partisan politics of Nicaragua were not the only hazard facing Alemán, who was also conscious of the need to conform to the expectations and condition- alities imposed by international financial institutions. His biggest concern was maintaining the approval of the International Monetary Fund:

The fact that the government hesitated greatly before even declaring a state of emergency after Mitch is evidence of a desire to avoid the responsibility for allocating massive resources to emergency assistance, thus increasing public spending and violating the conditions imposed by the structural adjustment programme. Another possible reason for not declaring a state of emergency was that a failure to mobilise large-scale human resources for the relief effort would (and did) expose the extremely limited capacity to respond to such situations by the scaled-down civil service. (Rocha and Christoplos, 2001: 249)

Indeed, what led to Alemán's political downfall and eventual conviction on multiple counts of corruption was not the loss of popular support, but the loss of support from international financial institutions. In Nicaragua authoritarianism is the norm and corruption is extremely high, but these issues have typically remained low on the list of popular concerns (IDESO, 2001). President Alemán's top-down and personalised approach to governance and his stunningly high level of corruption before and after Mitch (Walker, 2000) failed to significantly change pre-existing popular opinion of regime legitimacy. However, carrying out the Washington consensus required that the leadership maintain international legitimacy. When this was lost and it became clear that international players favoured Enrique Bolaños it set off a chain reaction as Alemán's support network realigned itself to accept the new internationally approved leader, and abandoned Alemán to his fate. A change in president had been affected, but this served to strengthen the neo-liberal orientation of government so that ideological regime change was not achieved.

Human security has not prospered under neo-liberalism. Effective risk management has not met expectations post-Mitch in El Salvador or Nicaragua (Wisner, 2000) despite receiving international aid and adopting a national rhetoric of 'learning the lessons of Mitch', neoliberal state restructuring has precluded their implementation. Comfort *et al.* (1999) use the Honduran and Nicaraguan cases to support their argument that risk and hazard mitigation strategies should be integrated into social policy, especially development schemes. They point to ways in which the shift in the social contract under neoliberalism from local to global interests is given material expression through land use and ultimately the distributions of risk in society.

Hurricane Mitch was remarkable because of the tremendous loss of life, social upheaval and economic devastation that it wrought but also because of the various local, national, regional, international and especially supranational responses it engendered. The restructuring of Central American states to conform to the United States' political and economic agenda for the post-Cold War period, which is sometimes referred to as the Washington Consensus, did not of course begin with Hurricane Mitch. However, this disaster gave the United States, the World Bank and International Monetary Fund, and the Inter-American Development Bank a platform from which to further their visions for region-wide transformation. Indeed the name adopted by the group of advisors was the Consultative Group for the Reconstruction and Transformation of Central America, and its logo states that 'reconstruction must not be at the expense of transformation'. The second meeting of this group, held in Stockholm 25–8 May 1999, resulted in the 'Stockholm Declaration' in which six goals were elaborated:

- Reduce the social and ecological vulnerability of the region, as the overriding goal.
- Reconstruct and transform Central America on the basis of an integrated approach of transparency and good governance.
- Consolidate democracy and good governance, reinforcing the process of decentralisation of governmental functions and powers, with the active participation of civil society.
- Promote respect for human rights as a permanent objective. The promotion of equality between men and women, the rights of children, of ethnic groups and other minorities should be given special attention.
- Coordinate donor efforts, guide by priorities set by the recipient countries.
- Intensify efforts to reduce the external debt burden of the region.

These were exciting times. Disaster management – in other words, proactive, integrated climate change adaptation – had been explicitly tied to the goals of inclusive governance, decentralised power, citizen participation, the promotion of human rights and debt reduction. A number if assessments of progress have, unfortunately, not found these goals to have been met (Christoplos *et al.*, 2009). Even shortly after the declaration Rocha and Christoplos (2001) observe that while some conceptual advances were made at the national level, such as the recognition that more appropriate agricultural practices and soil conservation may mitigate future disasters, and that environmental concerns such as forest fires and extensive clear cutting for cattle ranching must be addressed, not all post-Mitch initiatives were working to reduce risk. For example, the World Bank proposed to the Ministry of Agriculture and Forestry that they establish a publicly financed scheme of 'rainfall insurance', which would reimburse farmers' losses during times of drought or flooding:

> The economic justification for such a programme is to encourage farmers to adopt higher risk strategies. Agro-ecological risk-reduction practices, such as inter-cropping, are described as obstacles to achieving maximum potential production. The argument is that if farmers knew that they would be reimbursed for losses they would take the risk of abandoning agro-ecological production techniques in order to obtain greater profit. (Rocha and Christoplos, 2001: 243)

As with other areas of political life, discursive competition was decisive in shaping policy. Feeding the Stockholm Declaration goals through the dominant neo-liberal lens had some curious but perhaps not surprising results. In the name of improving state performance, while some advances were made in citizen participation and corruption, more marked was support for large scale privatisation of state assets including sale of the state-owned telephone company, the restructuring of the electrical company, the private administration of water supply and the opening of the petroleum sector to private investors (IADB, 1999).

Rocha and Christoplos (2001) conclude that any impact on policies or their implementation is doubtful. This is for structural reasons that extend beyond the state and the influence of international institutions: 'The isolation of NGOs, as elite institutions with little base in true civil society, must be broken if they are to become a vehicle for the integration of disaster mitigation and preparedness concerns in national policies and institutional practice' (Rocha and Christoplos, 2001: 250).

Demeritt *et al.* (2005) argue that international financial institutions fostered government policy in post-Mitch Nicaragua have promoted the liberal modernising of the state through the legal creation of a disaster prevention system (SINAPRED). This structurally links virtually all state ministries, while at the same time privatising key state services converting Nicaragua into a lucrative site for transnational disaster prevention agencies. The continuing transnationalisation of security in Nicaragua has resulted in the creation of a cohort of high level bureaucrats who are ostensibly responsible for designing and orchestrating a national disaster prevention plan; however, their primary work consists of middle-managing foreign projects and their work product is monitored and evaluated by World Bank officials, not the Nicaraguan people (Demeritt *et al.*, 2005).

2005, New Orleans, USA: transformation denied by political dilution

Hurricane Katrina has become a touchstone event in the USA. It demonstrates well the deep social, and in this case racial, determinants of individual adaptive capacity and vulnerability and the ways in which successive administrative systems failed first to mediate in the urban development that generated exposure and then to respond to the disaster and recovery. The disaster brought scrutiny and questions of legitimacy for local and national politicians, leading to reform in technical and administrative systems but limited indication of transformational reform. The lack of transformation is not for want of alternative visions or discourses but the failure of these to overcome popular rejection of politics and the dilution of political power through privatisation, which simultaneously removes public accountability (see below).

Michael Dyson (2006) found racial inequity in New Orleans an important enough factor in the Katrina disaster to begin his book with a discussion of some of the social conditions that reproduce it:

> New Orleans has a 40 percent literacy rate: over 50 percent of black ninth graders will not graduate in four years . . . Louisiana expends an average of $4,724 per student and has the third-lowest rank for teacher salaries in the nation. The black dropout rates are high and nearly 50,000 students cut class every day. When they are done with school, many young black males end up at Angola Prison, a correctional facility located on a former plantation where inmates still perform manual labor and where 90% of them will eventually die. New Orleans's employment picture is equally gloomy, since industry long ago deserted the city, leaving in its place a service economy that caters to tourists and that thrives on low-paying, transient and unstable jobs. (Dyson, 2006:8)

Part of the shame of New Orleans that cast doubt on political judgements was its predictability. In September 2004, just shy of one year before Hurricane Katrina, Jon Elliston published an article entitled 'Disaster in the Making'. The article carefully details how within months of gaining the presidency, George W. Bush presided over the dismantling and privatisation of the FEMA. Critical programmes developed over years were dropped, the agency's budget slashed and by 2004 FEMA had lost much of its capacity to fund mitigation projects. Meanwhile, following the 11 September 2001 attacks, the burden put on cash-strapped states to pay for anti-terrorism projects made them increasingly dependent on FEMA for help with disaster mitigation. But by 2004 the agency was forced to turn many away:

> In North Carolina, a state regularly damaged by hurricanes and floods, FEMA recently refused the state's request to buy backup generators for emergency support facilities. And the budget cuts have halved the funding for a mitigation program that saved an estimated $8.8 million in recovery costs in three eastern N.C. communities alone after 1999's Hurricane Floyd. In Louisiana, another state vulnerable to hurricanes, requests for flood mitigation funds were rejected by FEMA this summer. (Elliston, 2004)

After failing to win the bid for a flood mitigation project for Jefferson Parish (which one year later would represent together with New Orleans city proper 89 per cent of Katrina-affected population in the metropolitan area) (The Brookings Institution, 2005), Flood Zone Manager Tom Rodriguez told Elliston, 'You would think we would get maximum consideration for the funds. This is what the grant program called for. We were more than qualified for it' (Elliston, 2004).

Elliston's interviews with FEMA employees as well as unaffiliated academics reveal that the 'consultant culture' of privatisation had gutted what used to be a highly effective national service. States and communities now had to bid for mitigation grants from a diminished fund and in a system that made it harder for less affluent cities and communities to compete. Privatisation eroded the agency's institutional memory, effectively disregarding years of agency experience as disaffected staff joined the ranks of consultants. But as both scholars and practitioners observed, the lowest bidder does not necessarily do the best job, and private consultants do not necessarily accumulate and convert generations of experience into institutional memories that support effective action. In an essay submitted to the *New Yorker*, John McPhee (1987) couldn't have made the connection between business interests and maladaptive development any clearer:

> In the nineteen-fifties, after Louisiana had been made nervous by the St. Lawrence Seaway, the Corps of Engineers built the Mississippi River-Gulf Outlet, a shipping canal that saves forty miles by traversing marsh country straight from New Orleans to the Gulf. The canal is known as Mr. Go, and shipping has largely ignored it. Mr. Go, having eroded laterally for twenty-five years, is as much as three times its original width. It has devastated twenty-four thousand acres of wetlands, replacing them with open water. A mile of marsh will reduce a coastal-storm-surge wave by about one inch. Where fifty miles of marsh are gone, fifty inches of additional water will inevitably surge. The Corps has been obliged to deal with this fact by completing the ring of levees around New Orleans, thus creating New Avignon, a walled medieval city accessed by an instate that jumps over the walls. (McPhee, 1987)

The Army Corp of Engineers has been severely criticised for its lack of understanding of ecological systems, leading them to engage in counterproductive development and mitigation work. Nevertheless, it should be said that as engineers they recognised that the levee system might not hold up against a category four or five hurricane or even a category three if it hovered over the city. Indeed, one year before Katrina, the Corp proposed to study how New Orleans could be protected from a powerful hurricane, but according to independent journalist Sidney Blumenthal (2005), the Bush administration ordered that the research not be undertaken. Blumenthal moves up the chain of command to identify the administration as responsible for the policy that all but guaranteed disaster:

> The Bush administration's policy of turning over wetlands to developers almost certainly also contributed to the heightened level of the storm surge ... Bush had promised 'no net loss' of wetlands, a policy launched by his father's administration and bolstered by President Clinton. But he reversed his approach in 2003, unleashing the developers. The Army Corps of Engineers and the Environmental Protection Agency then announced they

could no longer protect wetlands unless they were somehow related to interstate commerce. (Blumenthal, 2005)

Against this backdrop of maladaptation in the early morning hours of 29 August 2005, meteorologists tracking the trajectory of category four Hurricane Katrina reported that it had shifted direction away from the city of New Orleans and was heading into the Gulf of Mexico. Many thought that a major disaster had been avoided. Then reports came in that some of the levees protecting New Orleans had been breached and that vast areas of the city were flooding. Soon afterwards, televised images began to appear. Though the storm affected a wide swath of Gulf Coast and killed at least 1,300 people in Louisiana, Mississippi, Alabama and Florida, Americans were transfixed by the spectacle of a humanitarian disaster unfolding in a major city in the United States.

The videos of mostly African-Americans struggling to survive in the hellish conditions of New Orleans' Superdome and Convention Center, in sweltering heat, with no electricity or running water, while basic supplies and transport failed to arrive, were juxtaposed in American consciousness to those seen a day earlier: tens of thousands of cars leaving the relatively white suburbs towards safety. As reports came in that the elderly residents of several nursing homes had also been abandoned, an already alarmed nation unused to confronting inequality in such stark relief ignited a national level search for blame (Frymer *et al.*, 2005).

Shock alone has not been enough to dislodge dominant cultural attitudes towards race and class in the US. Indeed some elements of the television media in particular have been criticised for resorting to presenting the disaster through a lens of cultural stereotypes that moved close to blaming the victims of the disaster for their own vulnerability (The Brookings Institution, 2005). Moreover, interpretations of the government's response to the crisis fractured along clear racial lines. According to a *Washington Post*-ABC poll, nearly three out of four white Americans did not believe that the government would have responded more quickly if the citizens trapped in the Superdome were wealthier and white, whereas the same proportion of blacks disagreed; and more than six in ten African-Americans believed that the poor relief effort reflected continuing racial inequity, while seven in ten whites rejected this view (Fletcher and Morin, 2005).

A second narrative in post-Katrina critique focused on administrative incompetence and had an overtly political dimension. The Democratic Party galvanised partisan anger through a mailer and email blitz focusing on the administration's incompetence; 'throw the bums out' urged the on-line organisation MoveOn.org. The mass media fuelled the campaign by first attacking Michael Brown, the head of FEMA, characterising him as a Bush crony who was awarded the job despite, it was claimed, being unqualified for the position. Even republicans wanted to know why the federal government was so slow to act.

Michael Brown's resignation shortly after the debacle caused a temporary lull in public furor. But this was revived in early March when the Associated Press distributed a videotape of Federal Disaster officials warning the president that the storm could breach levees, and recorded Brown voicing his concern that there were not enough disaster teams to help evacuees at the Superdome. The recorded briefing occurred one day before the hurricane hit. The videotape, along with seven days of briefing transcripts, raised doubts about the administration's claim that the 'fog of war' blinded them to the magnitude of the disaster, and directly contradicted the president's statement made four days after the storm, 'I don't think anybody anticipated the breach of the levees.' (Fletcher and Morin, 2005)

A more nuanced discourse has arisen from academic and think tank commentary that has tended to view the disaster from the viewpoint of those involved. One of two post-Katrina reports commissioned by The Brookings Institution (The Brookings Institution, 2005) demonstrates how the once racially mixed and vibrant city of New Orleans was transformed after World War II, and how these changes affected the contours of the 2004 disaster. Jim Crow laws, deindustrialisation and white flight into suburban neighbourhoods, combined with a host of federal housing and absurdly dangerous growth and land use policies. The result was the creation of a deeply segregated population of highland dwelling middle-class whites and increasingly marginalised lowland dwelling African-Americans, racially distinct communities existing in two different universes, with the latter living on borrowed land and borrowed time. A report from the Institute for Women's Policy Research observes that the establishment of three historically African-American colleges in the city supported the development of a highly educated African-American middle class, and that recovery projects should tap into this extremely valuable resource (Gault *et al.*, 2005). Another Brookings report is careful to observe that some middle-class whites suffered great losses, and not all of the poor are African-American (Berube and Katz, 2005). Yet all of these reports demonstrate that race, gender and poverty contributed separately and in combination with environmentally unsustainable policies to create particular citizens and communities significantly more vulnerable to hazards and crisis than others. They propose a radically different vision for the reconstruction of the metropolitan area, and offer concrete suggestions on how to integrate socio-economic diversity with ecological sustainability.

Douglas Brinkley's (2006) tome on the Katrina disaster provides historical depth and presents an impressive breadth of knowledge of the socio-political, cultural and environmental factors leading up to the disaster. Though most actors are presented as people caught up in larger processes beyond their full understanding or control, there is no shortage of villains in this narrative. Yet he reserves special rancour for the mayor of New Orleans, C. Ray Nagin. Making painfully clear the racial tenor of the critique, he observes that the pro-business mayor has long been called an 'Uncle Tom' by his detractors. Brinkley asserts that the reason Nagin failed to order a mandatory evacuation of New Orleans, when it would have saved lives and prevented thousands from suffering, was because he was afraid to anger his patrons in the business community whose wealth is maintained with the profits from tourism.

As the post Katrina political repercussions continue to unfold, the partisan battles become increasingly shrill, and the mainstream press dutifully reproduces the battles for public consumption, it remains to be seen if the voices that link disaster to wider socio-economic and environmental policies that increase vulnerabilities and reduce human security, for some more than others, will be heard over the din. (Brinkley, 2006: 23)

Despite the range of critical discourses following Katrina, tangible adaptive responses remain constrained. Birkland, in a study of policy learning following disaster events, undertook a review of the Library of Congress's Thomas database two months after the disaster. He found:

Of the 293 items this search returned, 40 percent of the bills mentioned Hurricane Katrina in the title, 24 percent included the work 'relief' in the title, and the items 'recovery' and 'reconstruction' were mentioned in 9 and 5 percent of titles, respectively. The word 'preparedness' appeared in three bills (1 percent) and the word [hazard] 'mitigation' did not appear in any bill. Clearly [hazard] mitigation or even preparedness was not a major concern of Congress in the two months after the disaster. (Birkland, 2007: 178)

Why is it that an event with demonstrable impact on popular, political and legislative consciousness failed to translate into progressive, proactive forms of adaptation built on hazard mitigation and disaster preparedness? Birkland (2007) argues that this was a result of the lack of an organised advocacy lobby around hurricane risk management, in turn a product of 'confusion over what it takes to improve policy performance and of political constraints that prevent officials from adopting effective policies' (Birkland, 2007: 178). He argues that, in the US, incremental learning has brought improvements in risk regulation such as the Flood Insurance Reform Act (transitional adaptation) but that the political leadership required for more transformational acts such as the banning of home construction on the coastline which would need a rethinking of development policy remains absent; an absence made all the more stark when compared to the impacts of the homeland security agenda, which indeed has taken resource and political attention away from hurricane and other disaster risk management. This is perhaps also explained by the modest levels of popular engagement with direct democracy in the US. Levels of trust in government and the governance of basic needs and security provision were dented by Hurricane Katrina but already low (Nicholls, 2009) and a sense that it is the system itself that is rotten rather than individual politicians or even parties, as one local newspaper commentator suggests:

Some political 'experts' believe that the recent hurricanes will create a demand for bigger government and a return to a New Deal-style of economics. This view is rooted in events that took place after the 1927 Mississippi River flood. But unlike the 2005 storms, government played a small role in the storm of 1927. Businessmen in New Orleans, who sacrificed St. Bernard and Plaquemines Parishes to save their own interests, were viewed as the villains. (Mainly because they never followed through with their commitment to reimburse the people of St. Bernard and Plaquemines Parishes for the damages they incurred.) In the aftermath of Hurricane Katrina the 'villains' appear to be the inflexible and uncaring bureaucracy of FEMA, and the indecisiveness and often bickering local political officials, and the many problems related to the administration of the Road Home program. In stark contrast to the storm of 1927, many individuals and corporations were clearly willing and able to help those affected by providing food, shelter, and even jobs. Therefore, we see Hurricane Katrina creating further disgust against the 'business as usual' politics of Louisiana. History has shown us that political shenanigans are least tolerated in times of suffering and/or blatant corruption. (PoliticsLA.com, no date)

As significant for some will be the impact of demographic changes wrought by Katrina with the residents of predominantly poor, black neighbourhoods being displaced through evacuation and redevelopment. A final reminder of the resilience of incumbent politicians in the US and especially New Orleans comes from the 1966 floods of the city which were also precipitated by poor government planning and levee failure during a hurricane. The incumbent mayor used disaster relief to bolster his public image and was re-elected to office a month later despite being personally responsible for the reallocation of city funds originally destined to shore up the levee (Abney and Hill, 1966).

Conclusion

What can these case histories tell us about adaptation and transformation? Certainly they show that decision-making for risk management is not only a technical matter but framed by and invested in the rough and tumble of politics. Importantly this is not only politics in the narrow sense of elections but rather more broadly refers to governance: the balance of private sector, state and voluntary sector provision of goods and services including the upholding of human rights. In each case this balance, which lies at the heart of the social contract, was challenged by disaster, and in particular by failures in the established regime to both reduce risk and prevent avoidable losses as part of everyday practice before the event, and be seen to act justly and effectively in response and reconstruction.

The most dramatic political outcome, the only example of regime change, unfolded in East Pakistan, when electoral politics failed to meet popular expectations for change. This is a prime example of disaster acting as a catalyst for political trajectories already entrained pre-disaster. The succession movement successfully presented the disaster as a symbol for state failure and provided the organisational base for propelling popular discontent into the political arena.

Table 8.1 Lessons for adaptation

	East Pakistan (Bangladesh)	Nicaragua	New Orleans, USA
Social contract	Pre-disaster contract lacked cultural legitimacy	Social contract held in place by international financial actors	Popular denial of inequalities in social contract
Disaster impact	Slow and limited state response fuelled succession movement	Opened regime to international scrutiny and pressure	Revealed chronic administrative failures at all levels of governance
Human security	In the long-run, basic needs and human rights were enhanced following independence	Lost opportunity to build capacity through governance reform	Evacuation, speculation and redevelopment have changed the demography of the city; the city remains at risk
Lessons for adaptation	Electoral politics is sensitive to popular discontent but when not heeded armed violence can result	International agendas for reform need to be built on local capacities and willingness to change	Overreliance on the private sector caused gross inefficiency, loss of knowledge and human resource in the public sector

The symbolic significance of disaster and failure in response and reconstruction is shown in each case. Following Mitch international actors sought to open Nicaragua's political-economy to the values of inclusive democracy. But battles lost at the level of discourse allowed distortions to appear in policy that undermined this agenda. The reinterpretation of decentralised governance as privatisation by international financial institutions is a case in point. Reflecting the scale of shock generated by Hurricane Katrina, this event has led to multiple competing discourses, there was though only marginal impact in the delivery of response. Given the administrative failures that prefigured the disaster one might expect a greater degree of internal reflection but this was not clearly present in the federal regime so that the impacts of the disaster were in many cases exacerbated by the exploitation of reconstruction opportunities by private contractors, large and small. The lack of an advocacy coalition and possibly also deep rooted popular disengagement from politics at the time weakened oversight and reflexivity.

Table 8.1 summarises each case study. Context is critical in judging the capacity for adaptation to open scope for progressive transformation. Adaptation can be championed through instrumental policy but also as part of a more diffuse cultural reaction to risk. It is at the conjuncture of cultural and instrumental adaptations that discourse is most influential, shaping the transfer of popular complaint into political prioritising and policy. Technical analysis is part, but not the whole, or even necessarily a dominant element, of this process of discursive competition which encompasses emotional as well as intellectual reasoning at a societal level. The narratives and metaphors that give collective shape to individual emotions hold the power to challenge or reinforce the status quo as adaptation to climate change becomes internalised as a social process – part of history as well as a technical and policy domain.

Part IV

Adapting with climate change

9 Conclusion: adapting *with* climate change

> Too frequently adaptation still reflects a narrow framing, which assumes that climate change is an ultimate, rather than a proximate driver of change.
>
> (Nelson, 2009: 496)

The potential, and even likely, implications of climate change for ecological and physical systems are profound and disturbing. Social systems that deliver specific management functions and organise governance serve to mediate between these impacts and people at risk. In this way understanding adaptive capacity and action requires a lens that can examine organisational behaviour and governance regimes, as well as the feelings, values and actions of individuals. Perhaps most important are the interactions between different levels of social actor (individuals and organisations) and the institutions that give shape to social systems. Research and policy on adaptation to climate change is just beginning to recognise the full contribution of values and governance to behaviour and action. Work on adaptation is emerging from an early period in its evolution as an intellectual domain where adaptation has, as Nelson (2009) rightly observes, been narrowly framed. Until now the overriding need has been for an articulation of adaptation as a function of climate change impacts (and for some a sub-set of vulnerability). Under the influence of the UNFCCC and IPCC this has in turn required but not quite achieved a clear definition. Adaptation, though, has in the process been separated from mitigation and development. As we have seen throughout this book, climate change is affecting socio-ecological systems in many ways. The majority are compound and indirect, and many quite ambiguous, so that it is difficult to imagine science will ever be able to identify the proportion of an expected or past event that is attributable to climate change alone and so precisely what climate change adaptation, narrowly defined, should be.

As our technical understanding of climate change adaptation is accompanied by a more nuanced view that can include governance as a field of adaptation, as well as a context within which technical adaptations unfold, so the relationship between humanity and climate change shifts. Sites for adaptation become internalised within socio-ecological systems. We turn from adapting *to* climate change, towards adapting *with* climate change. This is quite a leap and also requires an admittance that anthropogenic climate change is with us now, and is likely to be with us for the foreseeable future. Adapting *with* sees climate change as internal (a product of humanity's values, decisions and actions), but also its coevolution with the environment, so that neither environmental nor social change is independent (Castree and Braun, 2001). Once the social, and governance in particular, is given more emphasis so the opportunities arising from adaptation to enhance sustainable development become more apparent, the aim of adaptation becomes not one of defensive but progressive risk management as part of a renewed sustainable development. To be sure, the efficiency based engineering and economic debates that have tended to dominate technological adaptation thus far will remain central to the material expressions of adaptation. But they will now be positioned as part of a wider social and political agenda, so that understanding our capacity for adaptation and which adaptive options are preferable becomes a social, cultural and political as well as a technical or economic judgement.

The following discussion outlines the consequences of this perspective on adaptation for future research and policy and provides a synthesis of the argument and evidence for adapting *with* climate change made in this book.

How to adapt *with* climate change?

What are the consequences of moving from seeing climate change risk as an external threat to development to accepting that it is both a product and driver of development? Beck's seminal work on risk society provides some insight into the shape that future research and associated policy directions might take; in light of the proposals made in this book, two lessons can be taken.

First, groups in society compete not only for material wealth but also for security. In societies where wealth is achievable, security can become an overriding concern. Risk and its management is therefore a political as well as a technological concern and one where the poor are at risk of carrying a double burden when competition leads to their being marginalised from mechanisms providing security as well as wealth, or indeed where such burdens are interpreted as unfortunate but necessary costs for the production of security and wealth for others, elsewhere or in different times. This is a politic that needs to be critiqued from the perspective of social justice, leading to a vision of adaptation as a potential tool for progressive development, not one that uncritically defends the status quo.

Second, the risk society thesis contends that increasingly the most important risks are hard to detect and require technological innovation to make them visible. For climate change this is certainly the case and includes the challenge of making future risk and the carbon costs of risk reduction tangible and actionable in the present. The advances made in climate science have recently been applied to questions of adaptation through techniques like scenario planning that seek to provide decision-makers with a range of possible futures as a basis for planning infrastructure investments. This is useful but limited. Living with climate change means accepting future hazards cannot be planned out, or even necessarily predicted. Rather than seeking ever more precise technological guidance and solutions the urgency of climate change adaptation suggests we need to learn how to live with the fuzziness of climate change. Indeed what need to be made visible are not only the physical forcing mechanisms but also the human processes driving anthropocentric climate change and the distribution of its impacts. This lesson has already been learned at the sharp end of climate change adaptation practice. Here local vulnerability assessments place at least as much, if not more, emphasis on acting as a tool to facilitate local reflection on governance and underlying development processes as they do in providing technical accuracy on climate-change-associated vulnerability and risk (van Aalst *et al.*, 2008; Pelling, 2007b). The need to confront climate change risk by moving from a race for accuracy to a mechanism for studying governance as part of risk assessment is spreading, and has also been encompassed in methods for the participatory assessment of national capacity to manage disaster risk across the Americas by the InterAmerican Development Bank (IADB, 2005).

How might research and policy development on adaptation move forward? Building on the discussions made in this book four priorities for research are proposed that can help to better frame adaptation as a development problem:

Diversify the subject and object of adaptation research and policy

Early work on adaptation has rightly focused on a tightly bounded object for research and in so doing has succeeded in contributing to a clearly defined domain for policy. But if we see adaptation as a social as well as a technological phenomenon then there is a need to extend from this core. The object of analysis necessarily broadens from the behaviour of individuals and their constraining institutions to include organisations, governance systems, national and international politics. In parallel the subject of analysis extends from economy and technology to include cultural, social and political opportunities, play-offs and costs of adaptive options. Importantly it is in the interaction of different worldviews and priorities established from viewing adaptation through these contrasting lenses that the richness of adaptation policy, potential conflict and scope for coordinated and progressive, sustainable development could emerge.

Focus on social thresholds for progressive adaptation

Thresholds mark the tipping points from one systems state to another, and have been recognised in climate science and also through the concatenated impacts of climate change. Less work has been undertaken on thresholds between different stages of adaptation. Research on coping has long recognised the staggered nature of household responds to risk as economic pressures cause first non-productive and then productive assets to be expended and finally see the dissolution of households and migration as hazard impacts and vulnerability increase. The parsimony rule in cybernetics presents a similar guidance; that action requiring the least expenditure of resources will be undertaken first. But both coping and cybernetics focus on ex-post-adaptation; less is known about stages in proactive adaptation, which is curious given the volume of writing presenting this as the preferred adaptive form.

But focusing on a single adaptive choice or mechanism will be increasingly difficult, and miss the bigger picture of interactions between adaptations and the wider development agenda, as climate change impacts are felt through ever increasing multiple, direct and indirect pathways, often without being recognised. In this context critical thresholds will be those that set the broad scope for what is possible through adaptation and here the distinctions between resilience, transition and transformation are potentially helpful.

Recognise multiple adaptations: the vision effect

The interaction of multiple simultaneous adaptations has been recognised across scale when, for example, household adaptations are undermined or enhanced by local government action. But this is only one axis around which adaptation and efforts to shape adaptive capacity can interact. The competing values that underpin adaptation as resilience, transition and transformation indicate a 'vision effect' operating alongside the scale effect. This points to horizontal as well as vertical competition and complementarities in adaptation. This axis in large part explains the observed divergence between policy intention (policies) and emergence (self-organised activity) identified (Sotarauta and Srinivas, 2006) during the implementation of policy to support or enact adaptation; a gap that reveals tensions between the actions and values of competing adaptive strategies. The vision effect also helps explain difficulties in replicating, scaling-up and mainstreaming innovations that may be set within wider, contradictory visions of adaptation – local efforts at transformation will have most difficulty being mainstreamed if higher levels of governance construct adaptation as an act of resilience.

Link internal and external drivers of adaptation

Shifting of thinking on climate change from an external process to one unfolding as part of the coevolution of humanity and the environment makes it more important to understand internal – cognitive and cultural – drivers for adaptation. These are no longer fringe interests but part of the nexus of internal and external drivers that shape the who, where and when of adaptive capacity and action. The possibility that different adaptive initiatives could be in competition and lead to risk shifting between social groups and to non-human lives or future generations makes it all the more important to understand the deep psychological and cultural pressures that shape the propensity for different social groups to undertake particular adaptive strategies (including those that to the outside observer may appear to be self-limiting or detrimental to individual wellbeing).

Two final aspects of adaptation that researchers and policy-makers find especially difficult to grasp and that cross cut all of these emerging areas for policy and research are contingency and chance. What we do is no longer influenced only by local or even national processes and policies but also by increasingly unforeseen connections between systems, be they ecological, economic or political, worldwide. Scope for adaptation, as with any capacity, is exposed to such teleconnected linkages and this will bring surprises. Anticipating risk in this context becomes more difficult and consequently places greater emphasis on the core beliefs and capacities of a society – the generic attributes that can be applied to novel and unforeseen pressures. These lie in culture and governance, the roots of adaptation.

A synthesis of the argument

This penultimate section provides an overview and synthesis of the main discussion points made in the preceding chapters.

The age of adaptation

Climate change presents the early twenty-first century with a grand opportunity to reconfigure the meaning and trajectory of development. First mitigation and now adaptation provide global challenges that call for a rethinking of development goals, visions and methods. This is not the first time such an opportunity has arisen: in the 1980s and 1990s sustainable development presented dominant global and local political and economic systems with new challenges and promised to open space for progressive, international development. These opportunities have not yet been realised, and have rather been captured by and come to reinforce the established political-economy. Sustainable development has morphed into ecological modernisation. The faltering pace of international negotiations around climate change mitigation and adaptation indicate the enormity of the stakes and, if agreement can be reached, also the potential scope for revision.

But just how can adaptation open space for rethinking development? In looking forward to help answer this question Chapter 1 sketches the existing international intellectual and policy landscape within which reforms can take root. Though not presented in Chapter 1, Bangladesh is an early leader in state-sponsored adaptation planning that acknowledges the centrality of governance. With support from the UK's DFID, Bangladesh has proposed several technical programmes for adaptation. Each of these programmes is supported by a layer of social policy that at once indicates the social justice lying at the core of adaptation: distributional justice is supported through investment in a social protection scheme and procedural justice through partnership with community-based adaptation, though the extent to which governance reforms are implemented remains to be seen.

Importantly, these most fundamental arenas for adaptation are also targets for much ongoing development work by local communities in partnership with international development NGOs and humanitarian organisations so that a large proportion of what might be considered generic investment to build adaptive capacity is being undertaken now, but in an ad hoc way, without large scale collaboration. It is through coordination, as much as financial and technical support, that the emerging international architecture for adaptation can be made to contribute to the effectiveness of local actions to build capacity and adapt.

But as adaptation matures as a policy domain so its construction through the lens of leading international institutions like the IPCC and UNFCCC must also be revisited. The original imperative for the IPCC to mark out clearly what climate change adaptation might be, as an additional or separate act to mitigation and everyday development, is useful in policy terms, but in the long-run counter-productive. Adaptation on the ground is seldom an activity that can be neatly separated from others, making it difficult to single out support for activities that adapt to climate change. Accepting the cultural, social and political elements of adaptation only makes this more difficult. The solution proposed here is to move from adaptation defined only as a specific policy domain, to one that also accepts adaptation as an activity and aspiration that cross-cuts all development activities, so that we accept the reality of adapting with climate change. The provision of direct budgetary support instead of targeted development adaptation aid (that may well draw money from existing development budgets) is one practical step that supports this vision of adaptation. The result is that in the future adaptation may need to hold multiple definitions depending upon its application, in the same way that poverty is described in very technical terms for government poverty alleviation targets (for example, indicated by education, access to nutritional requirements, daily per capital income and so on), but also more broadly in the development of poverty alleviation programmes and local pro-poor practice (for example, livelihoods, wellbeing and entitlements). This raises a challenge of synthesis. But the worse risk is that adaptation is trapped as a technical concern and misses an opportunity to contribute to the rethinking of current unsustainable development visions and paths.

The adaptation tapestry

A wide variety of adaptive actions have been noted by the adaptation literature (see Smit *et al.*, 2000; Smit and Pilifosova, 2001). Chapter 2 identifies nine continuums along which individual acts of adaptation have been classified in the literature, according to the nature of the adaptive action (degree of collaboration, focus, forethought and phasing) or scope of impact (target, timescale, carbon awareness, social consequences, developmental orientation). Together these actions, potentially unfolding through different actors in response to the same climate change associated pressure or even on the same object, make for a rich adaptation tapestry.

Analysing the conditions that determine adaptive capacity and action and the coproduction of adaptation with risk and development is the core task for contemporary studies which build also on previous attempts to theorise adaptation. Four antecedents of contemporary adaptation studies are detailed in Chapter 2. Each offers lessons for contemporary work. Cybernetics, coevolution and adaptive management share roots in systems theory, a theoretical perspective shared by contemporary work on resilience that has come to influence adaptation thinking (Janssen *et al.*, 2006). Work on cybernetics offers caution for the systems approach in general which surfaces a tension between the imperatives of parsimony for individual adaptations (promoting a single best adaptation based on that which causes fewest resources to be expended) and the need for collective flexibility (adaptive capacity is enhanced by diversity). Watts (1983) further argues that the interpretation of systems theory in cybernetics makes it difficult to include values in analysis, and to consider the adaptive agent changing the system itself – excluding transition and transformation (see below) as adaptive possibilities under cybernetics (Morren, 1983). Coevolution (Norgaard, 1994, 1995) is especially useful to our argument because it provides a framework for placing adaptation within history, rather than seeing it as an end point in its own right, and also warns that adaptive actions can form critical junctures with no possibility of reversal.

The less abstract notions of adaptive management and coping also offer lessons for adaptation. Adaptive management has been designed to guide resource management with studies highlighting key challenges to the development of management systems where adaptive learning is built in (Walters, 1997; Medema *et al.*, 2008). These include perceived high costs in the short-term, discomfort at implications for credibility of managers that deliberately experiment in the knowledge some experiments will fail, and the difficulty in maintaining local stakeholder commitment over the medium timespan needed to follow and compare experiments. Coping has been explored through an extensive range of writing and policy in the last 30 years with considerable overlap and lessons for adaptation. For example, Burton *et al.* (1993) propose four periods in the escalation of adaptive action, which in turn can be used to signify tipping points in systems behaviour as thresholds into each stage are breached: the movement from risk absorption (it is not felt) to tolerance (it is felt but not acted upon); risk tolerance to risk reduction (risk management is implemented); and finally risk reduction to radical change where management practices are unable to cope and risk manifests as unacceptable and unpreventable loss. Each of these thresholds could be breached by increasing hazardousness, but also reduced adaptive capacity and increased vulnerability – for example, through demographic or economic change. The literature on coping also makes it clear that multiple actors will have viewpoints on what to protect, enhance or expend through adaptive actions and these may not be easily resolved, their origins being in values and beliefs so that a key challenge for adaptation in heterogeneous societies is to reveal these different values as a first step to inclusive planning for climate change adaptation.

The resilience–transition–transformation framework

The antecedents and current work on adaptation provide a rich basis for analysis, but they do not yet capture the full significance of adapting to climate change as a dynamic in socio-ecological coevolution. Besides the technical inefficiencies of overlapping adaptations that have been identified in the literature as a scale effect of adaptation, there are deeper political and even epistemological frictions to be identified and addressed in adaptation planning and research. It is here that the proposed framework of adaptation as resilience, transition and transformation aims to make its contribution. These three levels of adaptation stand as distinct categories of intention and action. The theory used to make each aspect visible and assist in analysis accumulates so that, for example, social learning and self-organisation (the core of resilience) can also be applied to help understand transitional and transformational adaptation. No one form of adaptation is preferable, with any judgement being dependent upon viewpoint and context. The aim of making these forms of adaptation visible is to surface the tensions between policies and actions aimed at maintaining the status quo or seeking broader change in relations of social and political power through adaptation.

Resilience

Drawing from socio-ecological systems theory (Gunderson and Holling, 2002; Folke, 2006), the IPCC identifies three attributes of resilient systems: functional persistence, self-organisation and social learning. From the perspective of adaptation, resilience is made distinct because of the aspiration of maintaining functional persistence. This can allow unsustainable or socially unjust practices to persist as well as protecting common goods (Jerneck and Olsson, 2008). Self-organisation (the ability of the components of a system to organise without formal, hierarchical direction) and social learning (the capacity for new values, ideas or practices to be disseminated, popularised and become dominant in society or a sub-set such as an organisation or local community) can be found across all forms of adaptation. Of these two, arguably, social learning is the most critical. Social learning is as important for transitional or transformational adaptation. It requires a high level of trust, a willingness to take risks in order to extend learning opportunities, the transparency required to test and challenge embedded values, active engagement with civil society and a high degree of citizen partici-pation. The advantages for social learning where there is close interaction between social actors is clear, with social learning and self-organisation reinforcing one another, so that a social system exhibiting rich capacity for social learning is also likely to have considerable scope for self-organisation.

Transition

Transition is an intermediary form of adaptation (see Chapter 4) that seeks to realise full rights under existing political and governance regimes. Where the gap between legal rights and their application is large, transition will align itself closely with transformational adaptation, requiring significant efforts to overcome entrenched vested interests in the status quo. Where governance regimes function fully this gap and the need to aspire for transitional adaptation will be absent. Most likely transition will be felt as a series of incremental adaptations as rights claims are asserted. As rights turn from *de jure* to *de facto* the effect is to open space for new rights to be won so that over time transformational change may be observed. Young (1999) describes this as a bargaining process with depth of change being distinguished between that which takes place at the level of rules for decision-making; or more profound change of the transformational kind that unfolds at the level of norms and principles (Krasner, 1983). Analysing potential for transitional adaptation places focus on examining the persistence of institutions over time as much as how they may be changed, and the role of actors in this. For example, Gunderson and Holling (2002) refer to rigidity traps where people and institutions try to resist change and persist with their current management and governance system despite a clear recognition that change is essential.

Literature on socio-technological transitions has recently been applied to climate change mitigation (Haxeltine and Seyfang, 2009) and offers scope for helping to understand where and why adaptive transitions can be found. Applying this literature to transitional adaptation as conceived here comes with a caveat – so far this literature does not distinguish adequately between transitional and transformational change. Both pathways for change are used, sometimes synonymously. The former is taken as a sub-set of the latter, with transitional change an aspect of transformation and not identified as a goal in itself (Jerneck and Olsson, 2008). This said, the frameworks emerging from socio-technological transitions remain useful and more positively serve to show the closeness between transition and political aspects of transformation which on the ground may be hard to distinguish. For example, Geels and Schot (2007) observe that new ideas or discourses emerge from local protected spaces but their dissemination and capacity to change established values and practices in the regime is often determined by the extent to which higher-level (for example, international) actors and institutions support change. This is relevant for transitional and transformational change.

Transformation

Chapter 5 argues that for adaptation to be transformative and progressive it must provide scope for the revision and reform or replacement of existing social contracts and the meaning of security and modes of development, as well as defending social gains already won. This is a call to tackle the causes of vulnerability at their roots. For adaptation to be concerned with changing the assumptions and structures of how we think about and organise development, it must address the causes rather than only the symptoms of vulnerability and risk. The social sciences offer many lenses with which to critique development and derive alternatives. Here we outline three theses that have clear relevance to climate change adaptation: risk society, the social contract and human security.

Beck's (1992) risk society thesis is a critique of the atomising and fragmenting nature of modernity. This has led to dominant modes of contemporary development that too easily produce and do not compensate for or seek to prevent complex environmental and social harm. Risk society is reproduced through established values and assumptions about development and wellbeing held by individuals as well as being institutionalised through the organisation of the market, government and industry. If this thesis is accepted, then adapting to the risks associated with climate change needs to confront the way individuals perceive the world and their place in it, as well as challenging the organisation of development. This is both daunting and empowering – signalling as it does that each of us is a site for adaptive scrutiny.

The social contract describes the prevailing balance of rights and responsibilities in society and may be held in place by legitimate government or the rule of force. The social contract is determined by the balance of power in society. Culture, identity and the control of knowledge through education are frequently identified as key for realising political change by political (Habermas, 1985) and educational (Freire, 1969) theorists. Perhaps the most pointed cases of challenges to the social contract are those following shocks – economic, political or environmental – that manifest failure in the social contract to provide security from disaster. When climate change is associated with extreme events, then it is the potential for disaster to destroy place as well as social life (Hewitt, 1997) that opens scope for new understandings of identity and social organisation and an alternative to established structures in the social contract.

The notion of human security provides some substance to this argument. It places emphasis on the responsibility of the state to facilitate the meeting of human rights and basic needs for its citizens, and so goes beyond a narrow state-centric security (Gasper, 2005). Work on disasters has shown the frequency with which alternative social organisation arises post-event, and also the effectiveness with which democratic-market- and authoritarian-state-centred regimes close down opposition, sometimes violently (Pelling and Dill, 2010). Examples of transformative and progressive regime changes have also been observed. They are most likely when a pre-existing alternative provides a discursive and organisational base with which to frame and disseminate a critique of the social contract (Albala-Bertrand, 1993).

Sites of adaptive action

Most empirical work on adaptation thus far has studied local communities of place, justified by the location specific qualities of climate impacts, especially those associated with natural hazards. But as the theoretical framework of resilience–transition–transformation explains, adaptation unfolds within all social contexts from the internal to the global. Three important but as yet seldom studied contexts are used in this book to illustrate the resilience–transition–transformation framework for climate change adaptation: the organisation, the city and the nation-state.

Following Wenger (2000), Chapter 3 argues that the bounded spaces offered by organisations provide especially useful contexts within which to study processes of social learning and self-organisation. Five pathways are proposed through which adaptive action can be undertaken by individuals or discrete subgroups within an organisation: agent-centred reflexive adaptation, agent-centred institutional modification, agent-centred resource management, agent-led external action and organisational external action. Only the latter two are visible from outside the organisation as it acts to change its operating environment, which can include interaction with other organisations to effect regime level change, or acting as intermediaries in the transfer of knowledge to allow adaptive action to be taken by the most appropriate organisations. Chapter 6 presents an analysis of a dairy farmer's NGO, Grasshoppers, and one of the UK state agencies that regulates and advises on dairy farming, the Environment Agency. Both are examples of good practice with organisational form, as observed, not being maintained at the expense of function. Both also demonstrate the interaction of canonical and shadow social systems as vehicles for social learning. In Grasshoppers these systems are intertwined and difficult to separate. The Environment Agency, with its more formal structure, has suppressed shadow system activity but this is still critical for those actors who know how to 'work the system'. In this way the imperatives of a public agency for transparency and efficiency are to some extent in tension with those for adaptation, which is enhanced by diversity and where formal observation is limited to allow for experimentation, even where this fails or runs counter to the objectives of the canonical system but meets local needs.

Any city is a social construction. Competing visions of the city are underlain by ideological, material and economic interests (Kohler and Chaves, 2003). The balance of power between such completing visions determines the priorities and actions of political actors and organisations affecting the city. This provides scope for examining the extent to which formal and legal rights are exercised, how far different interest groups – those on the margins of society, business interests and so on, are able to organise to defend and claim rights. As Chapter 4 argues, for climate change adaptation this will take in those with stakes in both risk management and development policy and practice. Chapter 7 assesses and compares governance regimes from four urban settlements in Mexico's rapidly urbanising Caribbean coastline state of Quintana Roo, and finds evidence for resilience, transitional and transformational adaptation. Resilience is indicated by efforts to maintain business-as-usual development paths including those of the private sector mainstream but also migrant labourers drawn to the state for work with little organisation or affiliation to Quintana Roo. Transition is demonstrated by those civil society organisations that exercise existing legal and governance rights to confront unsustainable development with successes in preventing a small number of coastal developments, but little success in extending participation from consultation to meaningful engagement with the views of local actors in formulating urban development plans. Transformation is least visible, but found in the promotion of fundamentally alternative forms of development built around strengthening citizens' self-worth and association with local places. This echoes Freire's call for critical consciousness as a prerequisite for informed social change.

Nation-states have largely been left out of discussions on adaptation, beyond their roles as aid clients or donors or as regulators to set the policy landscape for local actors to adapt. But the political space of the state is also a site for adaptation and of competition between different vested interests, their visions for the state and its social contract with citizens and other private actors in the future. Risk society, the social contact and human security provide a framework for analysing the influence of adaptation on social relations within states. Central to this is the extent to which legitimacy is maintained by political actors following disaster events and subsequent reflection on the production of and responses to risk and loss. Chapter 8 presents lessons from Cyclone Bohla in East Pakistan (Bangladesh), Hurricane Mitch in Nicaragua and Hurricane Katrina in the United States. Each of these events had local impacts with national consequences – despite this only Hurricane Bohla is associated with regime change, in this case secession from West Pakistan. Hurricane Mitch stimulated progressive discourse at the international and local levels but this was not translated into action, with some reform agendas being reformulated through the lens of neoliberal restructuring – decentralisation became privatisation, for example. Following Hurricane Katrina, a number of discourses offered challenges to the Bush Administration and no doubt the disaster contributed to regime change at the following national elections, but the most profound impact seems to have been a deepening distrust in the political process. The high degree of private sector involvement in managing risk and reconstruction also served to distance the state from direct blame in this case and left citizens without a clear target for opposition.

From theory to action

The aim of this book has been to offer a constructive critique of the dominant trends in thinking about adaptation and climate change. Considerable progress has already been made in delineating a vision for adaptation that is amenable to the policy process. But such clarity as there is on adaptation is in danger of being won at the expense of tackling wider questions of development through the adaptation lens. In building a case for a deeper interaction between adaptation and development, material presented in this book has tried to stay close to the climate change adaptation debate. At the same time it has sought to broaden the debate by drawing on foundational social science works that point us in new directions for questioning what adapting to climate change should be for, and who should control the process.

In closing, Box 9.1 brings together the opening quotations from each chapter. Together they offer compelling 'highlights' of the adaptation story mapped out here, a story that has a long way to run.

From high beginnings framed by the Universal Declaration of Human Rights, Freire quickly reminds us of the challenges ahead for a progressive adaptation. Not only are external structures likely to resist change, but those at risk themselves are apt to choose to support and adapt to the status quo for lack of access to the tools and opportunities to develop and apply critical awareness. The IPCC formulation of adaptation to date aims to provide clarity for the policy community.

Box 9.1 Other voices make the case

Chapter 1: 'Everyone has the right to life, liberty, and security of person.' (Universal Declaration of Human Rights, Article 3)

Chapter 2: 'The adapted man, neither dialoguing nor participating, accommodates to conditions imposed upon him and thereby acquires an authoritarian and uncritical frame of mind.' (Paulo Freire, 1969: 24)

Chapter 3: 'The ability of a social or ecological system to absorb disturbances while retaining the same basic structure and ways of functioning, the capacity for self-organization, and the capacity to adapt to stress and change.' (IPCC, 2008: 880)

Chapter 4: 'When special efforts are made by a diffusion agency, it is possible to narrow, or at least prevent the widening of, socioeconomic gaps in a social system. In other words, widening gaps are not inevitable.' (Rogers, 1995, 442)

Chapter 5: 'Instead of destroying natural inequality, the fundamental compact substitutes, for such physical inequality as nature may have set up between men, an equality that is moral and legitimate, and that men, who may be unequal in strength or intelligence, become every one equal by convention and legal right.' (Rousseau, 1973, original 1762: 181)

Chapter 6: 'What matters is not structures, but relationships.' (Scientific advisor to the Welsh Assembly)

Chapter 7: 'In Cancun the most common idea is that "it is not my problem, if things go bad, I can flee to another state".' (Ex-member of the Quintana Roo State Congress)

Chapter 8: '. . . moments when underlying causes can come together in a brief window, a window ideally suited for mobilizing broader violence. But such events can also have extremely positive outcomes if the tension . . . are recognized and handled well.' (USAID, 2002)

Chapter 9: 'Too frequently adaptation still reflects a narrow framing, which assumes that climate change is an ultimate, rather than a proximate driver of change.' (Nelson, 2009: 496)

It does this well but should not be confused with a handbook for critical climate consciousness. In making its contribution the IPCC has stayed close to adaptation as resilience. In so doing this has so far bounded out much that can be achieved by transition and transformation. Amongst a range of social activists and thinkers, Rogers and Rousseau remind us of the need for critical consciousness to prevent

the loss of hard won social gains and for social progress to be at the heart
of development. Taken together, comments from those facing climate change
impacts, from a scientific advisor to the Welsh Assembly, an ex-member of the
Qunitana Roo State Congress to USAID, show the rich policy landscape of relevance to climate change adaptation and the need to mainstream policy and research
into the concerns of everyday development for any aspect of resilience, transition
or transformation to succeed. Finally, speaking from the climate change literature,
Nelson succinctly captures the framing challenge for climate change adaptation,
which the argument and framework presented in this book have sought to face.
Climate change is an expression of deeper and often harder to grasp socio-
ecological relationships. Adapting to climate change then requires strategies that
address these root causes as well as the more proximate concerns. The linkages
are there to be made – between livelihoods and governance, or choices on how to
spend and invest surplus wealth and connected value systems. We need to make
them soon.

References

van Aalst, M., Cannon, T. and Burton, I. (2008) Community Level Adaptation to Climate Change: The Potential Role Of Participatory Community Risk Assessment, *Global Environmental Change*, 18(1): 165–79.

Abney, G. and Hill, L. (1966) Natural Disasters as a Political Variable: The Effect of a Hurricane on an Urban Election, *American Political Science Review*, 60(4): 974–81.

Abrahamson, V., Wolf, J., Lorenzoni, I., Fenn, B., Kovats, S., Wilkinson, P., Adger, W. N. and Raine, R. (2009) Perceptions of Heatwave Risks to Health: Interview-based Study of Older People in London and Norwich, UK, *Journal of Public Health*, 31(1): 119–26.

Adams, W. M. (2008) *Green Development: Environment and Sustainability in a Developing World*, 3rd Edition, London: Routledge.

Adger, W. N. (2003) Social Capital, Collective Action and Adaptation to Climate Change, *Economic Geography*, 79(4): 387–404.

Adger, W. N. and Brooks, N. (2003) Does Global Environmental Change Cause Vulnerability to Disaster? in Pelling, M. (ed.) *Natural Disasters and Development in a Globalizing World*, London: Routledge.

Adger, W. N., Huq, S., Brown, K., Conway, D. and Hulme, M. (2003) Adaptation to Climate Change in the Developing World, *Progress in Development Studies*, 3 (3), 179–95.

Adger, W. N., Arnell, N. W. and Tompkins, E. L. (2005a) Successful Adaptation to Climate Change Across Scales, *Global Environmental Change*, 15, 77–86.

Adger, W. N., Brown, K. and Tompkins, E. L. (2005b) The Political Economy of Cross-scale Networks in Resource Co-management, *Ecology and Society*, 10:9. Accessed 10/05/2010 from http://www.ecologyandsociety.org/vol10/iss2/art9

Adger, W. N., Hughes, T. P., Folke Carpenter, S. R. and Rockstrom, J. (2005c) Social-ecological Resilience to Coastal Disasters, *Science*, 309: 1036–39.

Adger, W. N., Paavola, J., Huq, S. and Mace, M. J. (2006) Toward Justice in Adaptation to Climate Change, in Adger, N. W., Huq, S., Mace, M. J. and Paavola, J. (eds.) *Justice and Adaptation to Climate Change*. London: MIT Press.

Adger, W. N., Dessai, S., Goulden, M., Hulme, M., Lorenzoni, I., Nelson, D. R., Naess, L-O., Wolf, J. and Wreford, A. (2009a) Are There Social Limits to Adaptation to Climate Change? *Climatic Change*, 93, 335–54.

Adger, W. N., Eakin, H. and Winkels, A. (2009b) Nested and Teleconnected Vulnerabilities to Environmental Change, *Frontiers in Ecology and the Environment*, 7 (3), 150–58.

Adger, W. N., Lorenzoni, I. and O'Brien, K. L. (2009c) Adaptation Now, in Adger, W. N., Lorenzoni, I. and O'Brien, K. L. (eds.) *Adapting to Climate Change: Thresholds, Values, Governance*, Cambridge: Cambridge University Press.

Agrawala, S. (ed.), (2005) *Bridge Over Troubled Waters, Linking Climate Change and Development*, Paris: Organization for Economic Co-operation and Development.

Agyeman, J., Bullard, R. and Evans, B. (2003) *Just Sustainabilities: Development in an Unequal World*, London: Earthscan.

Albala-Bertrand, J. M. (1993) *Political Economy of Large Natural Disasters: With Special Reference to Developing Countries*, Oxford: Clarendon Press.

Alkire, S. (2003) *A Conceptual Framework for Human Security*, CRISE Working Paper #2, Queen Elizabeth House: University of Oxford. Accessed 10/05/2010 from http://www.crise.ox.ac.uk/pubs.shtml

Amador III, J. S. (2009), Community Building at the Time of Nargis: The ASEAN Response, *Journal of Current Southeast Asian Affairs*, 28(4): 3–22.

Amir, A. (no date) *What's New? Dawn the Internet*. Accessed 21/09/2009 from http://www.dawn.com/weekly/ayaz/20000512.htm

Anderies, J. M., Ryan, P. and Walker, B. (2006) Loss of Resilience, Crisis, and Institutional Change: Lessons from an Intensive Agricultural System in Southeastern Australia, *Ecosystems*, 9: 865–78.

Argyris, C. and Schön, D. (1978) *Organizational Learning: A Theory of Action Perspective*, London: Addison-Wesley.

Argyris, C. and Schön, D. (1996) *Organisational Learning II: Theory, Learning and Practice*, London: Addison-Wesley.

Armitage, D., Marschke, M. and Plummer, R. (2008) Adaptive Co-management and the Paradox of Learning, *Global Environmental Change*, 18(1): 86–98.

Arthur, W. B. (1989) Competing Technologies and Lock-in by Historical Events: The Dynamics of Allocation Under Increasing Returns, *Economic Journal*, 99: 116–31.

Atwell, R. C., Schulte, L. A. and Westphal, L. M. (2008) Linking Resilience Theory and Diffusion of Innovations Theory to Understand the Potential for Perennials in the U.S. Corn Belt, *Ecology and Society*, 14(1): 30. Accessed 10/05/2010 from http://www.ecologyandsociety.org/vol14/iss1/art30

Bateson, G. (1972) *Steps Towards an Ecology of the Mind*, New York: Ballentine.

Beck, U. (1992) *Risk Society: Towards a New Modernity*, London: Sage.

Beck, U. (1999) *World Risk Society*, Cambridge: Polity.

Beebe, N. W., Cooper, R. D., Mottram, P. and Sweeney, A. W. (2009) Australia's Dengue Risk Driven by Human Adaptation to Climate Change, *PLoS Neglected Tropical Diseases*, 3(5): e429. Accessed 10/05/2010 from http://www.ncbi.nlm.nih.gov/pmc/journals/532

Bendaña, Alejandro (1999) The Politics of Hurricane Mitch in Nicaragua, *The Post*, The Parkland Institute, December. Accessed 02/06/2009 from http://www.ualberta.ca/PARKLAND/post/Vol-III-No1/05bendana.html

Benight, C. C., Ironson, G., Klebe, K., Carver, C. S., Wynings, C. and Burnett, K. (1999) Conservation of Resources and Coping Self-efficacy Predicting Distress Following a Natural Disaster: A Causal Model Analysis Where the Environment Meets the Mind, *Anxiety, Stress, and Coping: An International Journal*, 12: 107–26.

Benson, J. K. (1977) Organizations: A Dialectic View, *Administrative Science Quarterly*, 22: 1–21.

Berkes, F. (2007) Understanding Uncertainty and Reducing Vulnerability: Lessons from Resilience Thinking, *Natural Hazards*, 41: 283–95.

Berkhaut, F., Smith, S. and Sterling, A. (2004) Socio-technological Regimes and Transition Contexts, in Elzen, B., Geels, F. W. and Green, K. (eds.) *System Innovation and the Transition to Sustainability: Theory, Evidence and Policy*, Cheltenham: Edward Elgar.

Berube, A. and Katz, B. (2005) Katrina's Window: Confronting Concentrated Poverty Across America, *Special Analysis in Metropolitan Policy*, The Brookings Institution.

Bicknell, J., Dodman, D. and Satterthwaite, D. (eds.) (2009) *Adapting Cities to Climate Change: Understanding and Addressing the Development Challenges*, London: Earthscan.

Bilgin, P. (2003) Individual and Societal Dimensions of Security, *International Studies Review*, 5, 203–22.

Birkland, T. A. (2007) *Lessons of Disaster: Policy Change After Catastrophic Events*, Washington DC: Georgetown University Press.

Birkmann, J. (2006) Measuring Vulnerability to Promote Disaster-resilient Societies: Conceptual Frameworks and Definitions, in Birkmann, J. (ed.) *Measuring Vulnerability to Natural Disasters: Towards Disaster Resilient Societies*, Hong Kong: UNU Press.

Le Blanc, D. (2009) Climate Change and Sustainable Development Revisited: Implementation Challenges, *Natural Resources Forum*, 33, 259–61.

Blumenthal, S. (2005) No One Can Say They Didn't See It Coming, *Online News*, 31 August. Accessed 04/07/2009 from http://dir.salon.com/story/opinion/blumenthal/2005/08/31/disaster_preparation/index.html

Booth, D. (ed.) (1994) *Rethinking Social Development: Theory, Research and Practice*, Harlow: Logman: Scientific and Technical.

Booth, K. (1991) Security and Emancipation, *Review of International Studies*, 17: 313–26.

Boyd, R. and Hunt, A. (2006) *Costing the Local and Regional Impacts of Climate Change Using the UKCIP Costing Methodology*, Metroeconomica Limited.

Brinkley, D. (2006) *The Great Deluge: Hurricane Katrina, New Orleans, and the Mississippi Gulf Coast*, New York: Harper Collins.

Brown, J. S. and Duguid, P. (1991) Organizational Learning and Communities of Practice: Towards a Unified View of Working, Learning and Innovation, *Organizational Science*, 2(1): 40–57.

Buckle, S. (1993) *Natural Law and the Theory of Property: Grotius to Hume*, Oxford: Clarendon.

Bulkeley, H. and Betsill, M. M. (2003) *Cities and Climate Change: Urban Sustainability and Global Environmental Governance*, London: Routledge.

Burton, I. (2004) *Climate Change and the Adaptation Deficit*, Occasional paper 1, Adaptation and Impacts Research Group, Toronto: Meteorological Service of Canada, Environment Canada. Republished in Schipper, E. L. and Burton, I. (eds.) (2009) *The Earthscan Reader on Adaptation to Climate Change*, London: Earthscan.

Burton, I., Kates, R. W. and White, G. F. (1993) *The Environment as Hazard*, New York: Guilford Press.

Burton, I., Challenger, G., Huq, S., Klein, R. and Yohe, G. (2007) *Adaptation to Climate Change in the Context of Sustainable Development and Equity*, IPCC Working Group II contribution to the Fourth Assessment Report, Cambridge: Cambridge University Press.

Campos Cámara, B. L. (2007) *Procesos de Urbanización y Turismo en Playa del Carmen, Quintana Roo*, Mexico: Plaza Y Valdés.

Caney, S. (2006) Cosmopolitan Justice, Rights and Global Climate Change, *Canadian Journal of Law and Jurisprudence*, XIX: 255–78.

Carr, L. (1932) Disaster and the Sequence-pattern Concept of Social Change, *The American Journal of Sociology*, 38(2): 207–18.

Carter, T. P., Marry, M. L., Harasawa, H. and Nishioka, N. (1994) *IPCC Technical Guidelines for Assessing Climate Change Impacts and Adaptations*, London: University College London Press.

Cash, D. W. and Moser, S. C. (2000) Linking Global and Local Scales: Designing Dynamic Assessment and Management Processes, *Global Environmental Change*, 10, 109–20.

Castells, M. (1997) *The Information Age: Economy, Society and Culture*, Oxford: Blackwell.

Castree, N. and Braun, B. (2001) *Social Nature: Theory, Practice and Politics*, London: Wiley-Blackwell.

Chambers, R. (1989) *Editorial Introduction: Vulnerability, Coping and Policy*, IDS *Bulletin*, 20(2): 1–7.

Christoplos, I., Rodríguez, T., Schipper, L., Alberto Narvaez, E., Bayres Mejia, K. M., Buitrago, R., Gómez, L. and Pérez, F. (2009) *Learning from Recovery after Hurricane Mitch: Summary Findings*. Accessed 10/05/2010 from http://www.prevention consortium.org

Church, C. (2005) *Sustainability: The Importance of Grassroots Initiatives*, paper presented at Grassroots Innovations for Sustainable Development Conference, UCL, London, 10 June. Accessed 10/05/2010 from http://www.uea.ac.uk/env/cserge/events/2005/grassroots/index.htm

Clarke, D. A. (2009) Adaptation, Poverty and Well-being: Some Issues and Observations with Special Reference to the Capability Approach and Development Studies, *Journal of Human Development and Capabilities*, 10(1): 21–42.

Cohen, M. D., March, J. G. and Olson, J. P. (1972) A Garbage Can Model of Organizational Choice, *Administrative Science Quarterly*, 17(5): 1–25.

Comenetz, J. and Caviedes, C. (2002) Climate Variability, Political Crises, and Historical Population Displacements in Ethiopia, *Environmental Hazards*, 4: 113–27.

Comfort, L., Wisner, B., Cutter, S., Pulwarty, R., Hewitt, K., Oliver-Smith, A., Wiener, J., Fordham, M., Peacock, W. and Krimgold, F. (1999) Reframing Disaster Policy: The Global Evolution of Vulnerable Communities, *Environmental Hazards*, 1: 39–44.

Costanza, R. (2003) A Vision of the Future of Science: Reintegrating the Study of Humans and the Rest of Nature, *Futures*, 35: 651–71.

Cuéntame . . . de Mexico. Accessed 20 July 2010 http://cuentame.inegi.gob.mx/mono grafis/informacion/qroo/default.aspx?tema=me&e=23

Cumming, G. S., Cumming, D. H. M. and Redman, C. L. (2006) Scale Mismatches in Socialecological Systems: Causes, Consequences, and Solutions, *Ecology and Society*, 11(1): 14. Accessed 10/05/2010 from http://www.ecologyandsociety.org/vol11/iss1/art14

Cuny, F. (1983) *Disasters and Development*, Oxford: Oxford University Press.

Cutter, S. L., Barnes, L., Berry, M., Burton, C., Evans, E., Tate, E. and Webb, J. (2008) A Place-based Model for Understanding Community Resilience to Natural Disasters, *Global Environmental Change*, 18: 598–606.

Davies, S. (1993) Are Coping Strategies a Cop Out, *Institute of Development Studies Bulletin*, 24(4): 60–72.

Demeritt, D., Dill, K., Webley, P. and Wooster, M. (2005) *Enhancing Volcanic Hazard Avoidance Capactiy in Central America Through Local Remote Sensing and Improved Risk Communication* (EVHAC), Terminal Report for DFID KaR Project R8181.

DFID (2004a) *Disaster Risk Reduction: A Development Concern*, London: DFID.

DFID (2004b) *Climate Change and Poverty: Making Development Resilient to Climate Change*, London: DFID.

Diamond, J. (2005) *Collapse: How Societies Choose to Fail or Succeed*, London: Penguin.

Diduck, A., Bankes, N., Clark, D., and Armitage, D. (2005) Unpacking Social Learning in Social-ecological Systems: Case Studies of Polar Bear and Narwhal Management in Northern Canada, in Berkes, F., Huebert, R., Fast, H., Manseau, M. and Diduck, A. (eds.), *Breaking Ice: Renewable Resource and Ocean Management in the Canadian North*, Calgary: Arctic Institute of North America and University of Calgary Press, pp. 269–90.

Drury, A. C. and Olson, R. S. (1998) Disasters and Political Unrest: an Empirical Investigation, *Journal of Contingencies and Crisis Management*, 6(3): 153–61.

Duffield, M. (2007) *Development, Security and Unending War: Governing the World of Peoples*, Cambridge: Polity Press.

Dyson, M. E. (2006) *Come Hell or High Water*, New York: Basic Civitas Books.

Eakin, H. C. and Wehbe, M. B. (2009) Linking Local Vulnerability to System Sustainability in a Resilience Framework: Two Cases from Latin America, *Climatic Change*, 93: 355–77.

Elliston, J. (2004) Disaster in the Making, *Independent Weekly*, 22 September. Accessed 19/06/2009 from http://www.indyweek.com

FAO (2008) *Crop Prospects and Food Situation Report*, #2, April. Accessed 05/06/2009 from ftp://ftp.fao.org/docrep/fao/010/ai465e/ai465e00.pdf

Fankhauser, S. (1998) *The Costs of Adapting to Climate Change*, Global Environment Facility (GEF) Working Paper No. 16, Washington, D.C: GEF.

Farrington, J. and Bebbington, A. (1993) *Reluctant Partners: Non-governmental Organisations, the State and Sustainable Agricultural Development*, London: Routledge.

Fernandez-Gimenez, M. E., Ballard, H. L. and Sturtevant, V. E. (2008) Adaptive Management and Social Learning in Collaborative and Community-based Monitoring: A Study of Five Community-based Forestry Organizations in the Western USA, *Ecology and Society* 13(2): 4. Accessed 10/05/2010 from http://www.ecologyand society.org/vol13/iss2/art4

Fletcher, M. and Morin, R. (2005) Bush's Approval Rating Drops to New Low in Wake of Storm, *Washington Post Online*, 13 September. Accessed 13/07/2009 from http://www.washingtonpost.com

Flood, R. L. and Romm, N. R. A. (1996) *Diversity Management: Triple Loop Learning*, London: Wiley.

Folke, C. (2006) Resilience: The Emergence of a Perspective for Social-ecological Systems Analyses, *Global Environmental Change*, 16(3): 253–67.

Foxton, T. J. (2007) Technological Lock-in and the Role of Innovation, in Atkinson, G., Dietz, S. and Neumayer, E. (eds.) *Handbook of Sustainable Development*, Cheltenham: Edward Elgar.

Freire, P. (1969, 2000) *Education for Critical Consciousness*, New York: Continuum.

Freire, P. (1970) *Pedagogy of the Oppressed*, New York: Continuum.

Frymer, P., Strolovitch, D. Z. and Warren, D. T. (2005) Katrina's Political Roots and Division: Race, Class and Federalism, in *American Politics, Understanding Katrina: Perspectives from the Social Sciences*, Social Science Research Council.

Füssel, H. M. (2007) Adaptation Planning for Climate Change: Concepts, Assessment Approaches and Key Lessons, *Sustainability Science*, 2(2): 265–75.

Gallopin, G. C. (2006) Linkages Between Vulnerability, Resilience, and Adaptive Capacity, *Global Environmental Change*, 16: 293–303.

Galtung, J. (1994) *Human Rights in Another Key*, Cambridge: Polity.

Gasper, D. (2005) Securing Humanity: Situating 'Human Security' as Concept and Discourse, *Journal of Human Development*, 6(2): 221–45.

Gault, B., Hartmann, H., Jones-DeWeever, A., Werschkul, M. and Williams, E. (2005) Poverty, Race, Gender and Class, in *The Women of New Orleans and the Gulf Coast: Multiple Disadvantages and Key Assets for Recovery*, Part 1, The Institute for Women's Policy Research.

Geels, F. W. (2005) Processes and Patterns in Transitions and System Innovations: Refining the Co-evolutionary Multi-level Perspective, *Technological Forecasting and Social Change*, 72(6): 681–96.

Geels, F. and Schot, J. (2007) Typology of Sociotechnical Transition Pathways, *Research Policy*, 36: 399–417.

Germanwatch (2008) *Climate Change Performance Index 2008*. Accessed 10/05/2010 from http://www.germanwatch.org/klima/ccpi.htm

Gierke, O. (1934) *Natural Law and the Theory of Society 1500–1800*, Cambridge: Cambridge University Press.

Glantz, M. (1976) *Politics of Natural Disaster*, London: Praeger.

Global Environmental Change and Human Security programme (GECHS) (2009) *Human Security*. Accessed 10/05/2010 from http://www.gechs.org/human-security

GNAW (2001) *Climate Change Wales: Learning to Live Differently*, report, Cardiff: Government of the National Assembly of Wales.

Goodhand, J., Hulme, D. and Lewer, N. (2000) Social Capital and the Political Economy of Violence: A Case Study of Sri Lanka, *Disasters*, 24(4): 390–406.

Gore, C. (1992) *Entitlement Relations and 'Unruly' Social Politics: a Comment on the Work of Amartya Sen*, Mimeo, Brighton: Institute of Development Studies.

Gramsci, A. (1975/1992) Prison Notebooks, Vol. 1 (J. A. Buttigieg and A. Callari, trans.), New York: Columbia University Press.

Grasso, M. (2008) *A Normative Justice-based (Ethical) Analysis of the Funding of Adaptation to Climate Change at the International Level*, unpublished PhD thesis, King's College London.

Grin J., Rotmans, J. and Schot, J. (2010) *Transitions to Sustainable Development: New Directions in the Study of Long-Term Transformative Change*, London: Routledge.

Gross, R. (1996) *Psychology: The Science of Mind and Behaviour*, London: Hodder and Stoughton Educational.

Grothmann, T. and Patt, A. (2005) Adaptive Capacity and Human Cognition: The Process of Individual Adaptation to Climate Change, *Global Environmental Change*, 15(3): 199–213.

Grundmann, R. (2007) Climate Change and Knowledge Politics, *Environmental Politics*, 16(3): 414–32.

Guha-Sapir, D., Hargitt, D. and Hoyois, P. (2004) *Thirty Years of Natural Disasters 1974–2003: the numbers*, Brussels: Presses Universitaires de Louvain.

Gunderson, L. H. and Holling, C. S. (eds.) (2002) *Panarchy: Understanding Transformations in Social-ecological Systems*. London: Island Press.

Habermas, J. (1976) *Legitimation Crisis*, translated by Thomas McCarthy, London: Heinman.

Habermas, J. (1985) *A Theory of Communicative Action*, translated by Thomas McCarthy, London: Beacon Press.

Handmer, J. W. and Dovers, S. R. (1996) A Typology of Resilience: Rethinking Institutions for Sustainable Development, *Organization and Environment*, 9(4): 482–511.

Hansen, J. (2007) *Scientific Reticence and Sea Level Rise*, Environmental Research Letters 2. Accessed 10/05/2010 from http://www.iop.org/EJ/article/1748–9326/2/2/024002/erl7_2_024002.html

Harrison, N. (2003) Good Governance: Complexity, Institutions, and Resilience. Accessed 10/05/2010 from http://sedac.ciesin.columbia.edu/openmtg/docs/Harrison.pdf

Harvey, D. (2010) *The Enigma of Capital*, London: Profile Books.

Hasenclever, C., Mayer, P. and Rittberger, V. (1997) *International Regimes*, Cambridge: Cambridge University Press.

Haxeltine, A. and Seyfang, G. (2009) *Transitions for the People: Theory and Practice of Transitions and Resilience in the UK's Transition Movement*, working paper 134, Tyndall Centre for Climate Change Research, University of East Anglia.

Hewitt, K. (1997) *Regions of Risk: Geographical Introduction to Disasters*, London: Longman.

Hewitt, K. (ed.) (1983) *Interpretations of Calamity*, London: Allen and Unwin.

Hite, K. (1996) The Formation and Transformation of Political Identity: Leaders of the Chilean Left, 1968–90, *Journal of Latin American Studies*, 28(2): 299–328.

Hoffmann S. (2006) *Chaos and Violence: What Globalization, Failed States, and Terrorism Mean for U.S. Foreign Policy*, New York: Rowman & Littlefield Publishers.

Holling, C. S. (1973) Resilience and Stability of Ecological Systems, *Annual Review of Ecology and Systematics*, 4: 1–23.

Holling, C. S. (1978) *Adaptive Environmental Assessment and Management*, Chichester: Wiley.

Hudson, B. (2003) *Justice in Risk Society*, London: Sage.

Hulme, M. (2009) *Why We Disagree About Climate Change*, Cambridge: Cambridge University Press.

Huq, S. (2008) Community Based Adaptation, *Tiempo*, 68: 28.

IADB (1999) *Central America After Hurricane Mitch: The Challenge of Turning a Disaster into an Opportunity*, Consultative Group for the Reconstruction and Transformation of Central America. Accessed 12/11/2008 from http://www.iadb.org/regions/re2/consultative_group/backgrounder.htm

IADB (2005) *Indicators of Disaster Risk and Risk Management*, Washington DC: IADB. Accessed 10/05/2010 from http://www.iadb.org/regions/re2/consultative_group/backgrounder3.htm

IDESO (2001) *Elecciones y cultura política en Nicaragua*, Managua: Universidad Centroamericana, Instituto de Encuestas y Sondeos de Opinión.

IFRC (2005) *World Disasters Report*, Geneva: IFRC.

IFRC (2010) *World Disasters Report*, Geneva: IFRC.

INETER (1998) *Las Lluvias del Siglo en Nicaragua*, Managua. Accessed 20/09/2009 from http://www.ineter.gob.ni

IPCC (2007) *Climate Change 2007: Impacts, Adaptation and Vulnerability*, Cambridge: Cambridge University Press.

IPCC (2008) *Glossary of Terms for Working Group II*. Accessed 10/05/2010 from http://www.ipcc.ch/pdf/glossary/ar4-wg2.pdf

ISDR (2004) *Living with Risk: A Global Review of Disaster Reduction Initiatives*, Volume II, Geneva: ISDR.

ISDR (2005) *Hyogo Framework for Action 2005–2015: Building the Resilience of Nations and Communities to Disasters*, Geneva: ISDR.

ISDR (2009) *Global Assessment Report on Disaster Risk Reduction*, Geneva: ISDR.

Ison, R. L., High, C., Blackmore, C. and Cerf, M. (2000) Theoretical Frameworks for Learning-based Approaches to Change in Industrialised-country Agricultures, in Cerf, M. and Gibbon, D. (eds.) *Cow Up A Tree: Knowing and Learning for Change in Agriculture: Case Studies from Industrialised Countries*, Paris: INRA.

Iwanciw, J. G. (2004) *Promoting Social Adaptation to Climate Change and Variability through Knowledge, Experiential and Co-learning Networks in Bolivia*, La Paz, Bolivia: ComunidAd. Accessed 10/05/2010 from http://www.lapaz.nur.edu

Janis, I. (1989) Groupthink: The Problems of Conformity, in Morgan, G. (ed.) *Creative Organization Theory*, London: Sage Publications.

Jannsen, M. A., Schoon, M. L. Ke, W. M. and Borner, K. (2006) Scholarly Networks on Resilience, Vulnerability and Adaptation within the Human Dimensions of Global Environmental Change. *Global Environmental Change*, 16: 240–52.

Janssen, M. A., Anderies, J. M. and Ostrom, E. (2007) Robustness of Social-ecological Systems to Spatial and Temporal Variability, *Society and Natural Resource*, 20: 307–22.

Jeffrey, P. and McIntosh, B. S. (2006) Description, Diagnosis, Prescription: A Critique of the Application of Co-evolutionary Models to Natural Resource Management, *Environmental Conservation*, 33(4): 281–93.

Jerneck A. and Olsson, L. (2008) Adaptation and the Poor: Development, Resilience and Transition, *Climate Policy*, 8(2): 170–82.

Jordan, B. (1985) *The State: Authority and Autonomy*, New York: Basil Blackwell.

Kane, S. M. and Yohe, G. (2007) Societal Adaptation to Climate Variability and Change: An Introduction, *Climatic Change*, 45:(1) 1–4.

Kasperson, R. E., Golding, D. and Kasperson, J. X. (2005) Risk, Trust and Democratic Theory, in Kasperson, J. X. and Kasperson, R. E. (eds.) *The Social Contours of Risk: Volume I*, London: Earthscan.

Kay, J. (1997) The Ecosystem Approach: Ecosystems as Complex Systems, in Murray, T. and Gallopinn, G. (eds.) *Proceedings of the First International Workshop of the CIAT-Guelph Project 'Integrated Conceptual Framework for Tropical Agroecosystem Research Based on Complex Systems Theories'*, Cali, Colombia: Centro Internacional de Agricultura Tropical.

Keller, E. J. (1992) Drought, War and the Politics of Famine in Ethiopia and Eritrea, *Journal of Modern African Studies,* 30: 609–24.

Kelly, P. M. and Adger, W. N. (2000) Theory and Practice in Assessing Vulnerability to Climate Change and Facilitating Adaptation, *Climate Change,* 47(4): 325–52.

Kessler, R. C., Galea, S., Jones, R. T. and Parker, H. A. (2006) Mental Illness and Suicidality after Hurricane Katrina, *World Health Organization Bulletin*, 84: 930–9.

Klein, N. (2007) *The Shock Doctrine*, London: Penguin.

Klein, N. (2008) In the Wake of Catastrophe Comes the Whiff of Unrest, *The Guardian,* 16 May, p. 35.

Klinenberg, E. (2002) *A Social Autopsy of Disaster in Chicago*, London: University of Chicago Press.

Klüver, J. (2002) *An Essay Concerning Sociocultural Evolution, Theoretical Principles and Mathematical Models*, Dordrecht, the Netherlands: Kluwer Academic Publishers.

Kohler, G. and Chaves, E. J. (eds.) (2003) *Globalization. Critical Perspectives*, New York: Nova Science Publishers.

Kolm, S. C. (1996) *Modern Theories of Justice*, Cambridge: The MIT Press.

Konow, J. (2003) Which Is the Fairest One of All? A Positive Analysis of Justice Theories, *Journal of Economic Literature*, XLI: 1188–1239.

Krankina, O. N., Dixon, R. K., Kirilenko, A. P. and Kobak, K. I. (1997) Global Climate Change Adaptation: Examples from Russian Boreal Forests, *Climatic Change*, 36: 197–215.

Krasner, S. D. (1982) Regimes and the Limits of Realism: Regimes as Autonomous Variables, *International Organization*, 36(2): 490–517.

Krasner, S. D. (1983) *International Regimes*, Ithaca: Cornell University Press.

Kristjanson. P., Reid, R. S., Dickson, N., Clark, W. C., Romney, D., Puskur, R., MacMillan, S., and Grace, D. (2009) Linking International Agricultural Research Knowledge with Action for Sustainable Development, *Proceedings of the National Academy of Sciences*, 106(13): 5047–52.

Kuhlicke, C. and Kruse, S. (2009) Ignorance and Resilience in Local Adaptation to Climate Change – Inconsistencies between Theory-Driven Recommendations and Empirical Findings in the Case of the 2002 Elbe Flood, *Gaia-ecological Perspectives for Science and Society*, 18(3): 247–54.

Labbate, G. (2008) The Incremental Cost Principle and the Conservation of Globally Important Habitats: a Critical Examination, *Ecological Economics*, 65(2): 216–24.

LaPorte, R. (1972) Pakistan in 1971: The Disintegration of a Nation, *Asian Survey*, 2: 97–108.

Lawson, E. J. and Thomas, C. (2007) Wading in the Waters: Spirituality and Older Black Katrina Survivors, *Journal of Health Care for the Poor and Underserved*, 18: 341–54.

Leach, M., Mearns, R. and Scoones, I. (1997) *Environmental Entitlements: A Framework for Understanding the Institutional Dynamics of Environmental Change*, discussion paper #359, Brighton: Institute of Development Studies, University of Sussex.

Lebel, L., Anderies, J. M., Campbell, B., Folke, C., Hatfield-Dodds, S., Hughes, T. P. and Wilsonet, J. (2006) Governance and the Capacity to Manage Resilience in Regional Social-ecological Systems. *Ecology and Society*, 11: 9. Accessed 10/05/2010 from http://www.ecologyandsociety.org/vol11/iss1/art19

Lee, E. K. O., Shen, C. and Tran, T. V. (2009) Coping with Hurricane Katrina: Psychological Distress and Resilience among African American Evacuees, *Journal of Black Psychology*, 35: 5–23.

Lee, K. N. (1993) *Compass and Gyroscope: Integrating Science and Politics for the Environment*, Washington DC: Island Press.

Leonard, L. (2009) *Civil Society Reflexiveness in an Industrial Risk Society: The Case of Durban*, South Africa, unpublished PhD thesis, King's College London.

Lewis, P. (1979) Axioms for Reading the Landscape: Some Guides for the American Scene, in Meining D (ed.) *The Interpretation of Ordinary Landscapes: Geographical Essays*, Oxford: Oxford University Press.

Linley, P. A. and Joseph, S. (2004) Positive Change Following Trauma and Adversity: A Review, *Journal of Traumatic Stress*, 17: 11–21.

Liu, J. Dietz, T., Carpenter, S. R., Alberti, M., Folke, C., Moran, E., Pell, A. N., Deadman, P., Kratz, T., Lubchenco, J., Ostrom, E., Ouyang, Z., Provencher, W., Redman, C. L. Schneider, S. H. and Taylor, W. W. (2007) Complexity of Coupled Human and Natural Systems, *Science*, 317: 1513–6.

López-Marrero, T. and Yarnal, B. (2010) Putting Adaptive Capacity into the Context of People's Lives: a Case Study of Two Flood-prone Communities in Puerto Rico, *Natural Hazards*, 52(2): 277–97.

Lorenzoni, I., Jordan, A., Hulme, M., Turner, R. K. and O'Riordan, T. (2000) A Co-evolutionary Approach to Climate Change Impact Assessment: Part I Integrating Socio-economic and Climate Change Scenarios, *Global Environmental Change*, 10: 57–68.

McEvoy, D., Lindley, S. and Handley, J. (2006) Adaptation and Mitigation in Urban Areas: Synergies and Conflict, *Municipal Engineer*, 159(4): 185–91.

MAFF (2000) *Climate Change and Agriculture in the United Kingdom*, London: Ministry of Agriculture Fishers and Food, HMSO.

McPhee, J. (1987) The Control of Nature: Atchafalaya, *The New Yorker*, 23 February. Accessed 09/07/2009 from http://www.newyorker.com

Manuel-Navarrete, D., Pelling, M. and Redclift, M. (2009) *Coping, Governance and Development. The Climate Change Triad*, EPD working paper #18. Accessed 10/05/2010 from http://www.kcl.ac.uk/schools/sspp/geography/research/epd/working.html

Manyena, S. B. (2006) The Concept of Resilience Revisited, *Disasters*, 30(4): 433–50.

Massey, D. (1994) *Space, Place and Gender*, Cambridge: Polity Press.

Maturana, H. and Varela, F. (1992) *The Tree of Knowledge – the Biological Roots of Human Understanding*, Boston: Shambala.

Medema, W., McIntosh, B. S. and Jeffrey, P. J. (2008) From Premise to Practice: A Critical Assessment of Integrated Water Resources Management and Adaptive Management Approaches in the Water Sector, *Ecology and Society*, 13(2): 29. Accessed 10/05/2010 from http://www.ecologyandsociety.org/vol13/iss2/art29

Melosi, V. M. (2000) Environmental Justice, Political Agenda Setting and the Myths of History, *Journal of Political History*, 12(1): 43–71.

Milly, P. C. D., Betancourt, J., Falkenmark, M., Hirsch, R. M., Kundzewicz, Z. W., Lettenmaier, D. P. and Stouffer, R. J. (2008) Stationarity is Dead: Whither Water Management? *Science*, 319(5863): 573–4.

MINDEF (Medical Mission to East Pakistan) (1970). Accessed 18/09/2009 from http://www.mindef.gov.sg/imindef/about_us/history/birth_of_saf/v03n11_history.html

Moench, M. (2007) Adapting to Climate Change and the Risks Associated with Other Natural Hazards. Methods for Moving from Concepts to Action, in Schipper, E. L. and Burton, I. (eds.) (2009) *The Earthscan Reader on Adaptation to Climate Change*, London: Earthscan.

Moos, R. H. (2002) The Mystery of Human Context and Coping: An Unravelling of Clues, *American Journal of Community Psychology*, 30(1): 67–88.

Morren, G. B. Jr. (1983) A General Approach to the Identification of Hazards and Responses, in Hewitt, K. (ed.) *Interpretations of Calamity*, London: Allen and Unwin.

Mungai, D. N., Ong, C. K., Kitame, B., Elkaduwa, W. and Sakthivadivel, R. (2004) Lessons from Two Long-term Hydrological Studies in Kenya and Sri Lanka, *Agriculture, Ecosystems and Environment*, 104: 135–43.

Nelson, D. R. (2009) Conclusions: Transforming the World, in Adger, W. N., Lorenzoni, I. and O'Briene, K. L. (eds.) *Adapting to Climate Change: Thresholds, Values, Governance*, Cambridge: Cambridge University Press, pp. 491–500.

Nelson, D. R., Adger, W. N. and Brown, K. (2007) Adaptation to Environmental Change: Contributions of a Resilience Framework, *Annual Review of Environment and Resources*, 32: 395–419.

Nelson, R. R. and Winter, S. G. (1982) *An Evolutionary Theory of Economic Change*, Cambridge, MA: Belknap Press.

Nicholls, K. (2009) *The Impact of Hurricane Katrina on Trust in Government*, paper presented at the Midwest Political Science Association 67th Annual National Conference. Accessed 15/10/2009 from http://www.allacademic.com/meta/p361444_index.html

Norgaard, R. B. (1994) *Development Betrayed: the End of Progress and a Coevolutionary Revisioning of the Future*, London: Routledge.

Norgaard, R. B. (1995) Beyond Materialism: A Coevolutionary Reinterpretation of the Environmental Crisis, *Review of Social Economy*, LIII(4): 475–92.

North, D. C. (1990) *Institutions, Institutional Change and Economic Performance*, Cambridge: Cambridge University Press.

O'Brien, K. (2009) Do Values Subjectively Define the Limits to Climate Change Adaptation? in Adger W. N., Lorenzoni, I. and O'Brien, K. L. (eds.) *Adapting to Climate Change: Thresholds, Values, Governance*, Cambridge: Cambridge University Press.

O'Brien, K. and Leichenko, R. M. (2003) Winners and Losers in the Context of Global Change, *Annals of the Association of American Geographers*, 93(1): 89–103.

O'Brien, K., Sygna, L., Leichenko, R., Adger, W. N., Barnett, J., Mitchell, T., Schipper, L., Tanner, T., Vogel, C. and Mortreux, C. (2008) *Disaster Risk Reduction, Climate Change and Human Security*, GECHS Report 2008:3, University of Oslo.

O'Brien, K., Hayward, B. and Berkes, F. (2009) Rethinking Social Contracts: Building Resilience in a Changing Climate, *Ecology and Society*, 14(2): 12. Accessed 10/05/2010 from http://www.ecologyandsociety.org/vol14/iss2/art12

Oldenburg, P. (1985), A Place Insufficiently Imagined: Language, Belief, and the Pakistan Crisis of 1971, *Journal of Asian Studies*, 44(4): 711–33.

Olson, R., Alvarez, R., Baird, B., Estrada, A., Gawronski, V. and Sarmiento Prieto, J. P. (2001) *The Storms of '98: Hurricanes Georges and Mitch – Impacts, Institutional Response, and Disaster Politics in Three Countries*, Special Publication #38, Natural Hazards Center, Institute of Behavioral Science, University of Colorado. Accessed 10/05/2010 from http://www.colorado.edu/hazards/publications/sp/sp38/sp38.html

Olson, R. S. and Gawronski, V. (2003) Disasters as 'Critical Junctures' Managua, Nicaragua 1972 and Mexico City 1985, *International Journal of Mass Emergencies and Disasters*, 21(1): 5–35.

Olsson, P., Folke, C. and Berkes, F. (2004) Adaptive Co-management for Building Resilience in Social-ecological systems, *Environmental Management*, 34: 75–90.

Olsson, P., Gunderson, L. H., Carpenter, S. R., Ryan, P., Lebel, L., Folke, C. and Holling, C. S. (2006) Shooting the Rapids: Navigating Transitions to Adaptive Governance of Social-ecological Systems, *Ecology and Society*, 11(1). Accessed 10/05/2010 from http://www.ecologyandsociety.org/vol11/iss1/art18

Osteen, M. (ed.) (2002) *The Question of the Gift: Essays Across Disciplines*, London: Routledge.

Ostrom, E. (2005) *Understanding Institutional Diversity*, Princeton, US: Princeton University Press.

Oxfam (2008) *Climate, Poverty and Justice*, Oxfam Briefing Paper #124. Accessed 17/02/2009 from http://www.oxfam.org.uk/resources/policy/climate_change/down loads/ bp124_climate_poverty_poznan.pdf

Paavola, J. (2005) Seeking Justice: International Environmental Governance and Climate Change, *Globalizations*, 2(3): 309–22.

Paavola, J. and Adger, W. N. (2006) Fair Adaptation to Climate Change, *Ecological Economics*, 56: 594–609.

Paavola, J., Adger, W. N. and Huq, S. (2006) Multifaceted Justice in Adaptation to Climate Change, in Adger, W. N., Paavola, J., Huq, S. and Mace, M. J. (eds.), *Fairness in Adaptation to Climate Change*, London: The MIT Press.

Pateman, C. and Mills, C. (2007) *Contract and Domination*, Cambridge: Polity.

Paton, D. and Johnston, D. (2006) *Disaster Resilience: An Integrated Approach*, Springfield: Charles C. Thomas.

Pelling, M. (2003a) Paradigms of Risk, in Pelling, M (ed.) *Natural Disasters and Development in a Globalizing World*, London: Routledge.

Pelling, M. (2003b) *The Vulnerability of Cities: Natural Disasters and Social Resilience*, London: Earthscan.

Pelling, M. (2007a) The Rio Earth Summit, in Potter, R. and Desai, V. (eds.) *The Arnold Companion to Development Studies*, London: Arnold.

Pelling, M. (2007b) Learning from Others: Scope and Challenges for Participatory Disaster Risk Assessment, *Disasters*, 31(4): 373–85.

Pelling, M. (2009) The Vulnerability of Cities to Disasters and Climate Change: A Conceptual Introduction, in Brauch, H. G. (ed.) *Coping with Global Environmental Change, Disasters and Security*, London: Springer.

Pelling, M. (2010) Hazards, risks and global patterns of urbanization, in Wisner, B., Kelman, I., and Gaillard, J. C. (eds.) *Routledge Handbook of Natural Hazards and Disaster Risk Reduction and Management*, London: Routledge.

Pelling, M. and Uitto, J. I. (2001) Small Island Developing States: Natural Disaster Vulnerability and Global Change, *Environmental Hazards: Global Environmental Change B*, 3: 49–62.

Pelling, M. and High, C. (2005) Understanding Adaptation: What Can Social Capital Offer Assessments of Adaptive Capacity? *Global Environmental Change*, 15(4): 308–19.

Pelling, M. and Dill, K. (2006) *Natural Disasters as Catalysts of Political Action*, ISP/NSC briefing paper 06/01, London: Chatham House.

Pelling, M. and Dill, K. (2010) Disaster Politics: Tipping Points for Change in the Adaptation of Socio-political Regimes, *Progress in Human Geography*, 34: 21–37.

Pelling, M., High, C., Dearing, J. and Smith, D. (2007) Shadow Spaces for Social Learning: A Relational Understanding of Adaptive Capacity to Climate Change within Organisations, *Environment and Planning A*, 40(4): 867–84.

Pelling, M. and Wisner, B. (eds.) (2009) *Disaster Risk Reduction: Cases from Urban Africa*, London: Earthscan.

Perrow, C. (1999) *Normal Accidents: Living With High-Risk Technologies*, Princeton: Princeton University Press.

Pisani, M. (2003) The Negative Impact of Structural Adjustment on Sectoral Earnings in Nicaragua, *Review of Radical Political Economics*, 35(2): 107–25.

Plummer, R. and Armitage, D. (2007) A Resilience-based Framework for Evaluating Adaptive Comanagement: Linking Ecology, Economics and Society in a Complex World, *Ecological Economics*, 61: 62–74.

PoliticsLA.com (no date) *The Political and Demographic Aftermath of Hurricane Katrina*. Accessed 15/10/2009 from http://www.politicsla.com

Poovey, M. (1998) *History of the Modern Fact: Problems of Knowledge in the Sciences of Wealth and Society*, Chicago: University of Chicago Press.

Pred, A. and Watts, M. (1992) *Reworking Modernity: Capitalism and Symbolic Discontent*, New Jersey: Princeton University Press.

Pritchard, L. J. and Sanderson, S. E. (2002) The Dynamics of Political Discourse in Seeking Sustainability, in Gunderson, L. H. and Holling, C. S. (eds.) *Panarchy: Understanding Transformations in Human and Natural Systems*, Washington, D.C.: Island Press, pp. 147–72.

Pugh, J. and Potter, R. B. (eds.) (2003) *Participatory and Communicative Planning in the Caribbean: Lessons from Practice*, Aldershot: Ashgate.

Putnam, R. (1993) *Making Democracy Work: Civic Traditions in Modern Italy*, Princeton: Princeton University Press.

Rapparport, R. (1967) *Pigs for the Ancestors: Ritual in the Ecology of a New Guinea People*, New Haven: Yale University Press.

Rawls, J. (1971) *A Theory of Justice*, Cambridge: Harvard University Press.

Reason, J. T. (1990a) The Contribution of Latent Human Failures to the Breakdown of Complex Systems, *Philosophical Transactions of the Royal Society of London B*, 37: 475–84.

Reason, J. T. (1990b) *Human Error*, Oxford: Oxford University Press.

Reason, J. T. (1997) *Managing the Risks of Organizational Accidents*, Aldershot: Ashgate.

Redclift, M. (1987) *Sustainable Development: Exploring the Contradictions*, London: Methuen.

Reeder, T., Wicks, J., Lovell, L. and Tarrant, O. (2009) Protecting London from Tidal Flooding: Limits to Engineering Adaptation, in Adger, W. N., Lorenzoni, I. and O'Brien, K. L. (eds.) *Adapting to Climate Change: Thresholds, Values, Governance*, Cambridge: Cambridge University Press.

Rip, A. and Kemp, R. (1998) Technological Change, in Rayner, S. and Malone, E. L. (eds.), *Human Choice and Climate Change*, vol 2, Columbus: Battelle Press.

Rocha, J. L. and Christoplos, I. (2001) Disaster Mitigation and Preparedness on the Nicaraguan Post-Mitch Agenda, *Disasters*, 25(3): 240–50.

Rodgers, M. (1999) *In Debt to Disaster: What Happened to Honduras After Hurricane Mitch?* London: Christian Aid.

Rogers, E. M. (1995) *Diffusion of Innovations*, New York: The Free Press.

van Rooy, A. (ed.) (1998) *Civil Society and the Aid Industry*, London: Earthscan.

Rose, A. (2004) Defining and Measuring Economic Resilience to Disasters, *Disaster Prevention and Management*, 13(4), 307–14.

Rotmans, J., Kemp, R. and van Asselt, M. (2001) More Evolution than Revolution: Transition Management in Public Policy, *Foresight*, 3(1): 15–31.

Rousseau, J. J. (1973, original 1762), *The Social Contract and Discourses*, London: Dent and Sons.

Saldaña-Zorrilla, S. O. (2008) Stakeholders' Views in Reducing Rural Vulnerability to Natural Disasters in Southern Mexico: Hazard Exposure and Coping and Adaptive Capacity, *Global Environmental Change*, 18(4): 583–97.

Satterthwaite, D., Huq, S., Reid, H., Pelling, M. and Romero-Lankao, P. (2009) Adapting to Climate Change in Urban Areas: The Possibilities and Constraints in Low- and Middle-income Nations, in Bicknell, J., Dodman, D. and Satterthwaite, D. (eds.) *Adapting Cities to Climate Change: Understanding and Addressing the Development Challenges*, London: Earthscan.

Schipper, E. L. and Pelling, M. (2006) Disaster Risk, Climate Change and International Development: Scope and Challenges for Integration, *Disasters*, 30(1): 19–38.

Schipper, E. L. and Burton, I. (2009) Understanding Adaptation: Origins, Concepts, Practice and Policy, in Schipper, E. L. and Butron (eds.) (2009) *The Earthscan Reader on Adaptation to Climate Change*, London: Earthscan.

Schneider, S. H. (2004) Abrupt Non-linear Climate Change, Irreversibility and Surprise, *Global Environmental Change*, 14: 245–58.

Schneider, S. H. and Londer, R. (1984) *Coevolution of Climate and Life*, San Fransisco: Sierra Club Books.

SDN (Sustainable Development Networking) (no date) Victory Day and Liberation War 1971: Background of Liberation War, United Nations Development Programme.

Sehgal, I. (2005) Coping With Disaster, Opinion, *The International News*, Internet Edition, 8 September. Accessed 02/09/2009 from http://www.thenews.com.pk/editorial_arc1. asp?wn=Ikram%20Sehgal&page=4

Sen, A. K. (1981) *Poverty and Famines*, Oxford: Oxford University Press.

Sen, A. K. (1987) *On Ethics and Economics*, Oxford: Basic Blackwell.

Sen, A. K. and Dreze, J. H. (1999) Democracy as a Universal Value, *Journal of Democracy*, 10: 3–17.

Sen, B. V. (1973) Moscow and the Birth of Bangladesh, *Asian Survey*, 13(5): 482–95.

Seo, M. G. and Creed, W. E. D. (2002) Institutional Contradictions, Praxis, and Institutional Change: A Dialectical Perspective, *Academy of Management Review*, 27(2): 222–47.

Seyfang, G. and Smith, A. (2007) Grassroots Innovations for Sustainable Development: Towards a New Research and Policy Agenda, *Environmental Politics*, 16(4): 584–603.

Sharer, R. J. (2006) *The Ancient Maya*, 6th Edition, Stanford: Stanford University Press.

Shaw, P. (1997) Intervening in the Shadow Systems of Organizations: Consulting from a Complexity Perspective, *Journal of Organizational Change*, 10(3): 235–50.

Shue, H. (1999) Global Environment and International Inequality, *International Affairs*, 75: 531–45.

Slobodkin, L. and Rappaport, A. (1974) An Optimal Strategy for Evolution, *Quarterly Review of Biology*, 49: 181–200.

Slovic, P. (1999) Perceived Risk, Trust and Democracy, in Cvetovich, G. and Löfstedt, R. (eds.) *Social Trust and the Management of Risk*, London: Earthscan.

Smit, B. (ed.) (1993) *Adaptation to Climatic Variability and Change*, Guelph: Environment Canada.

Smit, B. and Pilifosova, O. (2001) Adaptation to Climate Change in the Context of Sustainable Development and Equity, in McCarthy, J. J. (ed.) *Climate Change 2001: Impacts, Adaptation and Vulnerability, IPCCWorking Group II*, Cambridge: UK: Cambridge University Press, pp. 877–912.

Smit, B., Wandel, J. (2006) Adaptation, Adaptive Capacity and Vulnerability, *Global Environmental Change*, 16, 282–92.

Smit, B., Burton, I., Klein, R. J. T. and Wandel, J. (2000) An Anatomy of Adaptation to Climate Change and Variability, *Climatic Change*, 45: 223–51.

Smith, A. (2007) Translating Sustainabilities between Green Niches and Socio-technical Regimes, *Technology Analysis & Strategic Management*, 19(4): 427–50.

Smith, D. (1990) Beyond Contingency Planning – Towards a Model of Crisis Management, *Industrial Crisis Quarterly*, 4(4): 263–75.

Smith, D. (1995) The Dark Side of Excellence: Managing Strategic Failures, in Thompson, J. (ed.) *Handbook of Strategic Management*, London: Butterworth-Heinemann, pp. 161–91.

Smith, D. (2004) For Whom the Bell Tolls: Imagining Accidents and the Development of Crisis Simulation in Organizations, *Simulation and Gaming*, 35(3): 347–62.

Smith, J. B., Klein, R. J. T. and Huq, S. (eds.) (2003) *Climate Change, Adaptive Capacity and Development*, London: Imperial College Press. Accessed 10/05/2010 from ebooks at http://ebooks.worldscinet.com/ISBN/9781860945816/9781860945816.html

Smith, N. (1984) *Uneven Development: Nature, Capital and the Production of Space*, Georgia: University of Georgia Press.

Sommer, A. and Moseley, H. (1973) The Cyclone: Medical Assessment and Determination of Relief and Rehabilitation Requirements, in Chen, Lincoln (Ed), *Disaster in Bangladesh*, Oxford: Oxford University Press.

Sotarauta, M. and Srinivas, S. (2006) Co-evolutionary Policy Processes: Understanding Innovative Economies and Future Resilience, *Futures*, 38(3): 312–36.

Spence, P. R., Lachlan, K. A. and Burke, J. M. (2007) Adjusting to Uncertainty: Coping Strategies among the Displaced after Hurricane Katrina, *Sociological Spectrum*, 27: 653–78.

Splash, C. L. (2007) The Economics of Climate Change Impacts à la Stern: Novel and Nuanced or Rhetorically Restricted? *Ecological Economics*, 63(4): 706–13.

Stanton, E. A., Ackerman, F. and Kartha, S. (2008) *Inside the Integrated Assessment Models: Four Issues in Climate Economics*, Working Paper WP-US-0801, Stockholm Environment Institute. Accessed 10/05/2010 from http://devweb2.sei.se/us/Working Papers/WorkingPaperUS08-01.pdf

Stern, N. (2006) *The Stern Review on the Economics of Climate Change*, HM Government. Accessed 10/05/2010 from http://www.occ.gov.uk/activities/stern.htm

Susman, P., O'Keefe, P. and Wisner, B. (1983) Global Disasters, a Radical Interpretation, in Hewitt, K. (ed.) *Interpretations of Calamity*, London: Allen and Unwin.

Swift, J. (1989) Why Are Rural People Vulnerable to Famine? *IDS Bulletin*, 20(2): 815.

Tanner, T. and Mitchell, T. (2008) Entrenchment or Enhancement: Could Climate Change Adaptation Help Reduce Chronic Poverty? *IDS Bulletin*, 39(4): 6–15.

The Brookings Institution (2005) *New Orleans After the Storm: Lessons from the Past, a Plan for the Future*, Metropolitan Policy Program.

Tompkins, E. L. (2005) Planning for Climate Change in Small Islands: Insights from National Hurricane Preparedness in the Cayman Islands, *Global Environmental Change*, 15: 139–49.

Tompkins, E. L., Adger, W. N. and Brown, K. (2002) Institutional Networks for Inclusive Coastal Management in Trinidad and Tobago, *Environment and Planning A*, 34: 1095–111.

Torry, W. (1978) Natural Disaster, Social Structure and Change in Traditional Societies, *Journal of Asian and African Studies*, 13: 167–83.

Turner, B. A. (1976) The Organizational and Interorganizational Development of Disasters, *Administrative Science Quarterly*, 21: 378–97.

Turner, B. A. (1978) *Man-made Disasters*, London: Wykeham.

Tushman, M. and Anderson, P. (1986) Technological Discontinuities and Organization Environments, *Administrative Science Quarterly*, 31: 465–93.

UN-HABITAT (2007) *Global Report on Human Settlements, 2007: Enhancing Urban Safety and Security*, UN-HABITAT, London: Earthscan.

UN Human Rights Commission (2009) *Resolution 7/23, Human Rights and Climate Change*. Accessed 10/05/2010 from http://www2.ohchr.org/english/issues/climate change/docs/Resolution_7_23.pdf

UNDP (2004) *Reducing Disaster Risk: A Challenge for Development*, Geneva: UNDP.

UNDP (2007) *Fighting Climate Change: Human Solidarity in a Divided World, Human Development Report 2007/2008*, London: Palgrave.

UNFCCC (2007) *National Adaptation Programmes of Action*. Accessed 10/05/2010 from http://unfccc.int/national_reports/napa/items/2719.php

Unruh, G. C. (2000) Understanding Carbon Lock-in, *Energy Policy*, 28: 817–30.

Uphoff, N. (1993) Grassroots Organizations and NGOs in Rural Development: Opportunities with Diminishing States and Expanding Markets, *World Development*, 21(4): 607–22.

Urbinati, N. (1998) From the Periphery of Modernity: Antonio Gramsci's Theory of Subordination and Hegemony, *Political Theory*, 26(3): 370–91.

USAID (United States Agency for International Development) (2002) *Windows of Vulnerability and Opportunity. Foreign Aid in the National Interest: Promoting Freedom, Security and Opportunity.* Accessed 15/04/2009 from http://www.usaid.gov/fani/ch04/windows.htm

USAID (United States Agency for International Development) (2005) *Mission Accomplished: The United States Completes a $1 Billion Hurricane Relief and Reconstruction Program in Central America and the Caribbean.* Accessed 21/08/2009 from http://www.reliefweb.int/rw/RWB.NSF/db900SID/EGUA-6AYRX3?OpenDocument

USGS (United States Geological Survey) (1999) *Volcano Hazards Program, Pilot Project: Mount Rainier Volcano Lahar Warning System*, USGS.

Vale, L. J. and Campanella, T. J. (2005) *The Resilient City: How Modern Cities Recover from Disaster*, Oxford: Oxford University Press.

Vayda A. P. and McCay, B. (1975) New Directions in Ecology and Ecological Anthropology, *Annual Review of Anthropology*, 4: 293–306.

Vazquez, C., Cervellon, P., Perez-Sales, P., Vidales, D. and Gaborit, M. (2005) Positive Emotions in Earthquake Survivors in El Salvador (2001) *Journal of Anxiety Disorders*, 19: 313–28.

de Waal, A. (1997) *Famine Crimes: Politics and the Disaster Relief Industry in Africa*, Oxford: James Currey.

Waddell, H. (1975) How the Enga Cope with Frost: Responses to Climatic Perturbations in the Benral Highlands of New Guinea, *Human Ecology*, 3(4): 249–73.

Wagstaff, M. (2007) Remapping Regionalism, in Demossier, M. (ed.) *The European Puzzle: The Political Structuring of Cultural Identities as a Time of Transition*, Oxford: Berghahn Books.

Walker, B., Gunderson, L. H., Kinzig, A., Folke, C., Carpenter, S. and Schultz, L. (2006a). A Handful of Heuristics and Some Propositions for Understanding Resilience in Social-ecological Systems, *Ecology and Society*, 11: 13. Accessed 10/05/2010 from http://www.ecologyandsociety.org/vol11/iss1/art13

Walker, B., Salt, D. and Reid, W. (2006b) *Resilience Thinking: Sustaining People and Ecosystems in a Changing World*, Washington, D.C.: Island Press.

Walker, T. (2000), Nicaragua: Transition Through Revolution, in Walker, T. W. and Armony, A. C. (eds.). *Repression, Resistance and Democratic Transition in Central America*, Wilmington: Scholarly Resources, pp. 67–88.

Walsh, F. (2002) Bouncing Forward: Resilience in the Aftermath of September 11, *Family Process*, 41: 34–6.

Walters, C. J. (1997) Challenges in Adaptive Management of Riparian and Coastal Ecosystems, *Conservation Ecology*, 1(2): 1. Accessed 10/05/2010 from http://www.ecologyandsociety.org/vol1/iss2/art1

Walters, C. J. and Hilborn, R. (1978) Ecological Optimization and Adaptive Management, *Annual Review of Ecology and Systematics*, 9: 157–88.

Warner, J. (2003) Risk Regime Change and Political Entrepreneurship: River Management in the Netherlands and Bangladesh, in Pelling, M. (ed.) *Natural Disasters and Development in a Globalizing World*, London: Routledge.

Washington Post (1971) Disintegration and Opportunity in Pakistan, *Washington Post*, Washington, D.C., 17 March, p. A14.

Watts, M. (1983) On the Poverty of Theory: Natural Hazards Research in Context, in Hewitt, K. (ed.) *Interpretations of Calamity*, London: Allen and Unwin.

Watts, M. (2000) Political Ecology, in Sheppard, E. and Barnes, T. (eds.), *The Companion of Economic Geography*, Oxford: Blackwell, pp. 257–74.

Weinberg, A. M. (1972) Science and Trans-science, *Minerva*, 10: 209–22.

Wenger, E., (1999) *Communities of Practice: Learning, Meaning and Identity*, Cambridge: Cambridge University Press.

Wenger, E. (2000) Communities of Practice and Social Learning Systems, *Organization*, 7: 225–46.

White, A. L. (2007) *Is it Time to Rewrite the Social Contract?* Business for Social Responsibility, occasional paper. Accessed 10/05/2010 from http://www.tellus.org/publications/files/BSR_AW_Social-Contract.pdf

White, P., Pelling, M., Sen, K., Seddon, D., Russell, S. and Few, R. (2004) *Disaster Risk Reduction: A Development Concern*, London: DFID.

Williams, D. L. (2007) Ideas and Actuality in the Social Contract: Kant and Rousseau, *History of Political Thought*, 28(3): 469–95.

Williams, P. (2002) The Competent Boundary Spanner, *Public Administration*, 80(1): 103–124.

Willows, R. J. and Connell, R. K. (2003) *Climate Adaptation: Risk, Uncertainty and Decision-Making*, UKCIP Technical Report, London: UKCIP.

Wisner, B. (2000) Risk and the Neoliberal State: Why Post-Mitch Lessons Didn't Reduce El Salvador's Earthquake Losses, *Disasters*, 25(3): 251–68.

Wisner, B. (2006) *Let Our Children Teach Us: A Review of the Role of Education and Knowledge in Disaster Risk Reduction*, Geneva: ISDR and Action Aid.

Wisner, B., Blaikie, P., Cannon T. and Davis, I. (2004) *At Risk: Natural Hazards, People's Vulnerability and Disaster*, London: Routledge.

Yohe, G. and Tol, R. S. J. (2002) Indicators for Social and Economic Coping Capacity – Moving Toward a Working Definition of Adaptive Capacity, *Global Environmental Change*, 12, 25–40.

Yohe, G., Malone, E., Brenkert, A. Schlesinger, M. Meij, H. Xing, X. and Lee, D. (2006) *A Synthetic Assessment of the Global Distribution of Vulnerability to Climate Change from the IPCC Perspective that Reflects Exposure and Adaptive Capacity*, Palisades, New York: CIESIN, Columbia University. Accessed 17/02/2009 from http://ciesin.columbia.edu/data/climate

Young, O. R. (1999) *The Effectiveness of International Environmental Regimes: Causal Connections and Behavioural Mechanisms*, Cambridge: The MIT Press.

Young, O. R., Berkhout, F., Gallopin, G. C., Janssen, M. A., Ostrom, E. and van der Leeuw, S. (2006) The Globalization of Socio-ecological Systems: An Agenda for Scientific Research, *Global Environmental Change*, 16: 304–16.

Zadek, S. (2006) *The Logic of Collaborative Governance: Corporate Responsibility, Accountability and the Social Contract*, Working paper 17, Cambridge, Massachusetts: The Corporate Responsibility Initiative, John F. Kennedy School of Government, Harvard University. Accessed 10/05/2010 from http://www.hks.harvard.edu/m-rcbg/CSRI/publications/workingpaper_17_zadek.pdf

Zaman, M. Q. (1994) Ethnography of Disasters: Making Sense of Flood and Erosion in Bangladesh, *Eastern Anthropology*, 47: 129–55.

Zamani, Gh. H., Gorgievski-Dujvesteijn, M. J., Zarafshani, K. (2006) Coping with drought: Towards a Multilevel Understanding based on Conservation of Resources Theory, *Human Ecology*, 34(5), 677–92.

Zimmerer, K. S. and Basset, T. J. (eds.) (2003) *Political Ecology: An Integrative Approach to Geography and Environment-Development Studies*, New York: Guilford Publications.

Index